THE
HISTORY
OF ROTATION

THE HISTORY OF ROTATION

OF LIFE, WORLD, AND UNIVERSE

Louis Komzsik

THE REGENCY
PUBLISHERS

Library of Congress Control Number: 2022918300

ISBN: 978-1-959434-43-6   (Paperback Edition)
ISBN: 978-1-959434-44-3   (Hardcover Edition)
ISBN: 978-1-959434-42-9   (E-book Edition)

Book Ordering Information

The Regency Publishers, US
521 5th Ave 17th floor NY, NY10175

Phone Number: (315)537-3088 ext 1007
Email: info@theregencypublishers.com
www.theregencypublishers.com

Printed in the United States of America

For Leslie

CONTENTS

PROLOGUE

We live in a finite, small segment of an apparently infinite universe and we have long been fascinated by the forces and phenomena that appear to control it. The most intrinsic, and perhaps not generally recognized phenomenon in our universe, world and life is rotation. This phenomenon is pervasive in nature and permeates many facets of our life.

The book, as the title indicates, is mostly historical, however, at certain points a small bit of physics and mathematics is involved. The depth of these parts does not exceed the common high school level, but the reader troubled by them could skip those over and still enjoy the historical aspects of the book.

The invention and subsequent evolution of the mechanical wheel radically elevated humankind's ability to get around and carry objects. The recognition of the round nature of Earth, its rotational motion around the Sun giving us the years, and the rotation of Earth around its own axis giving us the days of our lives were the next recognized rotational effects.

The planetary levels of rotation were hard to accept for humans, we are unable to feel wind in our face due to the rotation of Earth, the most serious argument against it a couple of millennia ago. We certainly cannot feel the rotation of the solar system either, let alone that of the Milky Way, although its spiral shape, visible on the cover image, strongly suggests an influence of a rotational phenomenon. Nevertheless, we can prove those rotations either by observation or actual measurements. Earth's rotation gives rise to phenomena like hurricanes and trade winds, or even strangely us weighing more or less depending on the direction of our movement relative to Earth's.

Rotation also affects one of life's most enigmatic aspects, time. Celestial cycles were observed by ancient cultures and resulted in cosmic calendars. The large scale measuring of time was gradually refined by humankind with medium scale instruments of calendars that brought time to the horizon of a human lifetime. The smaller scale measurements, via rotational clocks of our everyday life, came next. Finally, the everyday manifestations of rotation in the very useful rotary machines, providing many conveniences of our lives, is also prevalent.

The phenomenon's universe wide interpretation leads to some intriguing philosophical contemplation including the question of the original cause of rotation.

1

WHEELING AROUND

About six thousand years ago the Egyptians moved very large blocks of stones on rollers created from tree trunks. It is accepted as the method of carrying large blocks of stone to great distances at other ancient monument building sites, such as Stonehenge. We do not know this for a fact, however, it is the most plausible by our present day knowledge of physics.

It is undeniable that recognition of the very first rotational phenomenon by our ancestors resulted in a tremendous acceleration of the evolution of man. Almost like evolution by revolution, the latter referring to the rotational effect, the subject of our interest, not the societal turmoil accompanying the other kind.

Considering the pace of approximately 2 million years of hominid history and about 200,000 years of homo sapiens' presence on Earth, the last ten thousand years' evolution is spectacularly accelerated. One cannot but wonder about the important contribution of the rotational phenomenon to this acceleration.

The rotational phenomenon has been utilized much earlier, albeit unrecognized as such. The well known ancient method of starting a fire, spinning a dry stick of wood between the palms with its tip drilling into another flat piece of dry wood, has survived the millenea. Some Amazonian native tribes, still disconnected from the mainstream of the human race and not exposed to modern tools, continue to rely on this method. It was probably the first manifestation of the influence of rotation in our life, while of course friction also played a role in generating the heat to spark the fire.

It was also about six thousand years ago that the actual wheel form, as we know it today, appeared in Mesopotamia. The Sumerians are credited with the invention, and its use enabled them to develop the first agricultural society. Of course, the first wheels were just round pieces of wood with a hole drilled in the center for the axle. To prevent the wheel from rolling off of the axle, a piece of metal, likely of bronze in those times, was fixed in place; the lynchpin as we know it today.

The wheel concept quickly found its way into other cultures. There is archaeological evidence of wheels in Europe and India about five thousand years ago, and in China about four thousand years ago. Its spreading like wildfire appears to be coincident with the domestication of cattle and horses at about the same time. Surely, the biggest benefit of the wheel came along with the invention of the cart pulled by a beast of burden.

Soon, humankind was wheeling around in cattle or horse drawn carriages all over the world, or even using it for entertainment as the Roman chariot races attested. The latter vehicles had more advanced wheels already; the invention of spokes allowed the wheels to be lighter but at the same time stronger. The spokes replaced the wheel's solid wood interior and its circumference was soon built from or at least reinforced by metal.

It is interesting to notice that the ancient American cultures, the Mayans, Incas and the Aztecs, were lagging behind in the invention of the wheel. It is in part likely due to their isolation and in the case of the Incas it may have also been related to their mountainous terrain that is not very conducive for wheeled travel. There are some archaeological finds of children's toys with wheel like components, but it is still believed that the practical use of wheel arrived in the Americas only in the 16th century with the Spanish conquistadors, incidentally along with the horse.

The next step in the wheel's evolution was even more instrumental in humankind's advancement. A deeper understanding of the rotational phenomenon led to the recognition of wheel's ability to do work. There are archaeological remnants of water wheels in Mesopotamia and Nubia where they were used for irrigation purposes. The water was scooped up when the buckets on the circumference of the wheel were in a lower position and when they were raised above the level of the axis of the wheel the water poured into a canal leading it to irrigate crops. It is a technology still in use today at some places on Earth where the rotation of the wheel is still provided by human or animal power.

Not long after that the reverse application also appeared. While scooping water from the moving flow of rivers, it was recognized that the flow can move the wheel on its own. Early Chinese evidence from 2-300 BC shows the usage of water power to grind grains. The wheel was mounted on an axle protruding from the side of a building and submerged into the water flow. On the inner end of the axle was a grinding stone that pulverized the grains on the top of a stationary stone basin.

The Romans developed this art into a science during the 1st century BC and AD. A historical anecdote is oft mentioned about the Romans abilities in the use of water wheels. According to the story, when the Visigoths surrounded Rome in the 6th century AD, they shut off the aqueducts supplying water to the city. The Romans suspended water wheels between two boats and anchored them to the pillars of the river Tiber where the flow was narrowed down and hence accelerated. The water lifted out by the water wheels supplied the city during the siege and the technique was proven so successful that it was adopted by other cities in medieval Europe.

The most notable water wheels were installed on the river Thames between the arches of the London Bridge. The plans called for the capability to pump more than one hundred thousand gallons of water per day to a height of 120 feet, a spectacular engineering feat for those times. The city fathers were, at first, hesitant to allow the construction, but the contractor demonstrated the power of the concept by shooting a water jet over the spire of a nearby church.

The contract was ultimately granted and the wheels were built in the second half of the 16th century. The distribution of the water was by the already existing water pipes, supplying the city for some hundred years from other sources. The wheels were destroyed in the great fire of London in 1666, but were later rebuilt and operated well into the 19th century. A truly spectacular longevity of a Roman invention making virtue out of necessity.

The influence of the first practical application of the rotating phenomenon did not stop there. It appears that once humans recognized the motion and its value, they were actively inventing more and more applications. The ancient engineers were thinking alike their modern successors. Fitting to human nature, the rotational phenomenon soon found its way into warfare. It was already in use in the form of the wheel in war chariots and carrying supplies for the troops, however, a radical step was made in the Greek city of Syracuse by the first mechanical engineer, Archimedes during the third century BC.

Archimedes was born in 287 BC and died defending his hometown in 212 BC. His personal life's details are not well known since his biography was lost, but his mathematical, physical and engineering accomplishments luckily survived the ages. His work in computing the volumes of spheres and cylinders, as well as his rule of objects floating in fluid are well known. It is his engineering work, however, where his understanding and use of the rotational phenomenon yielded the most spectacular results.

He used the concept of the lever and built a catapult equipment that threw stones and incendiary devices on the attacking Roman ships. He also developed a pulley system that enabled him to move large weights manually. The concept is now taught in elementary schools and is well understood that the force required to exert a certain pull is lessened by increasing the number of rope lines.

Specifically, the force required to lift an object with a two line pulley is one half of the weight of the object. Archimedes also realized that there is a price to pay for this advantage in the length of the pull. With such system the person needs to move the end of the rope by twice the distance of the object's motion. This is an early recognition of the conservation of the energy principle.

According to anecdotal evidence, Archimedes actually pulled a captured Roman ship onto the shore with his machinery single handed. His statement of "give me a place to stand and I will move the Earth" is part of recorded history and certainly testifies to his confidence in his physics.

Archimedes was one of the first scientists to investigate the properties of mathematical curves, called spirals. In fact, one of the spirals is called Archimedean spiral in modern mathematics. His spiral is described by a point moving away from its center with a constant radial velocity and simultaneously turning with a constant angular velocity. The mathematical spiral concept is aptly manifested in the arms of our Milky Way Galaxy shown on the cover image.

He also recognized the connection between the spiral shape and physical rotation, which led him to focus his attention to develop a water lifting device more advanced than the ancient water wheel. He placed a screw, just like the ones we use today but with a somewhat deeper thread, into a pipe with a tight fit that still enabled the screw to rotate. When the pipe was submerged into water and the screw turned, it carried the water up toward the top end. The water pouring out at the top was usually captured in another pipe and carried to the place of use, irrigated fields or bath houses.

That the Greeks had baths in his time was a fact commemorated by the anecdote of Archimedes' discovery of the facts of buoyancy. According to the story, upon discovering the rule, he jumped out of his bath and ran down naked on the main street of Syracuse yelling "Eureka" or "I found it".

The screw mechanism itself is really genial as it does not need precision in manufacturing. The surface of the screw is a simple helix, that is generated by wrapping the spiral around a cylinder. If the fit between the screw and the pipe is not tight enough, the water slipping back along the wall just ends up in the turn below and gradually makes its way up.

In a modified version of the design this issue does not even occur, the screw can be affixed solidly to the pipe wall and rotated together. The space between the pipe and the screw surface would gradually carry the water up.

This adaptation of Archimedes' screw was also used in its reverse form: water was poured in at the top making the pipe and the screw rotate. This rotation may have been connected with a tool and, there it was, the first incarnation of a generator, albeit a mechanical one.

Archimedes' understanding of the rotational phenomenon was far ahead of his time. He was also one of the early converts to the globular Earth concept, a topic of some controversy in ancient times.

2

CELESTIAL CYCLES

About 4,000 years ago the Mayan kingdom was founded in the Yucatan peninsula. Mayans are considered to be descendants of a wave of humans that crossed the Bering Strait around 10,000 years ago at the time of the last ice age. Their mythology describes a person of great abilities, who led their migration from the other side of the ocean on a mysterious open path. Ice age land bridge assumptions of modern archeology seem to agree with this ancient belief.

Their Yucatan kingdom founder, Kukulkan was considered to be the descendant of the Sun God. His organization of the nation into four tribes and their four tribal centers is supported by archaeological evidence. One of those centers, Chichen-Itza is substantially restored and is a testament to some of the Mayan achievements. Their kingdom lasted well into the middle ages and was ultimately destroyed in the middle of the 15th century by the Spanish colonists.

They were skilled astronomers, experts in observing the cycles in the sky. The Moon's cycle was an object of high importance. Because of the lack of artificial light, Moon had a big effect on everyday life by extending the daylight hours. Even on lower latitudes, like those of the Mayan territory where the daylight hours were longer and their seasonal change was less notable, the reliance on Moon was very strong.

Moon has continuously changing phases with very distinct shapes between its periods. The duration of these changes between two full Moon phases became the lunar cycle, the first and probably most influential celestial cycle. There was a less noticeable change between the full Moons, their size. Since Moon rotates around Earth on an elliptic orbit, the full Moon occurring when Moon is closest to Earth is larger than the one occurring when Moon is the farthest.

The result of this was that the lunar cycles changed and the Mayans noticed that. They even had a different hieroglyph for them. Interestingly they counted their lunar months in units of six, numbered from 0 to 5. Having a zero at such early period was remarkable, considering the fact that the advanced Mediterranean society of the Greeks still struggled with its recognition as a valid mathematical constant two millennia later.

The six lunar months may have been related to another celestial cycle that the ancient Mayans also intensely observed, the eclipses of Moon. They recognized a distinction between the two types of eclipses. The solar eclipse occurs when Moon passes in front of the Sun and when viewed from a certain region on Earth, completely covers it. The other eclipse type, the lunar eclipse occurs when Moon passes through the shadow of the Earth, which is visible from the night half of the Earth.

The probability of such a celestial event occurring is extremely low. Let us consider some numbers: The diameter of the Sun is approximately 864,000 miles and its average distance from us is about 93 million miles. Now the same numbers for the Moon are: diameter 2160 miles and its average distance is 238,000 miles, a stunning demonstration of the spectacular achievement of the Apollo astronauts landing on the Moon.

It is well known in everyday circumstances that an object at a distance appears smaller and smaller as the distance increases. Furthermore, a smaller object being closer than a larger one may appear the same size, a fact exploited by artists in medieval times. The apparent sizes of two objects are based on the ratio of their actual sizes and their distances. This is the concept at work, when we cover the Moon with our thumb in front of our eyes.

These numbers for our celestial objects are the following. The ratio of their sizes is 864000 / 2160 ≈ 400. The ratio ot their distances is 93000000 / 238000 ≈ 390. A most remarkable coincidence! The ratio by which the Sun is larger than the Moon is almost the same as the ratio by which the Moon is closer to Earth than the Sun, resulting in Moon's ability to cover the Sun during eclipses. The visual angle covering the Sun or the Moon are both about half a degree.

Because of the angle between the plane of the Moon orbiting Earth and the plane of the Earth's orbit around the Sun, the eclipses occur at irregular, albeit predictable times. Most of the time the Moon passes above or below the plane of Earth's orbit, and eclipses could only occur when Moon's orbit crosses the plane of Earth's orbit.

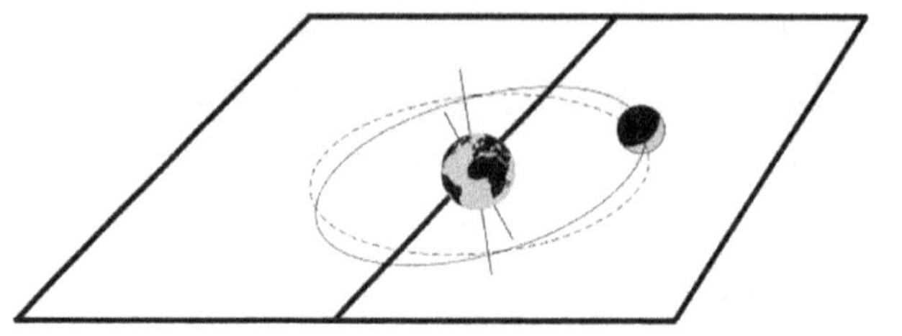 

Eclipses of Moon

Hence eclipses could only occur twice a year, approximately six months apart, that fact could also be the reason for the Mayans counting from zero to five. On the figure the solid circle represents Moon's orbit around the Earth, while the dotted line is the projection of Moon's orbit to Earth's orbital plane represented by the parallelogram on the figure.

It appears that the Mayans were experts at predicting eclipses, even though they may have not fully understood the reasons for the events. We have no records indicating any understanding of the solar system as rotating planets, or even the round Earth concept. Nevertheless, Moon's cycles and eclipses exerted huge influence on their society in general. Planting, harvesting and social events were aligned with Moon's phases and eclipses, according to rules set by the priests.

We know now that Moon is gradually getting farther away from us by almost two inches per year. It appears that Earth is slowly loosing gravitational control over its celestial companion. Four billion years ago, in Earth's youth, Moon was much closer and its influence much stronger causing volatile weather and huge, 40-60 feet high ocean waves. Moon still has a role on the tides in the oceans, an effect of Moon the Mayans were not aware of.

The Mayans were not only recording Moon's cycles, but those of Venus as well. They called Venus the Great Star, probably due to the fact that Venus was, and still is, the brightest celestial object next to Sun. Their recording of the Venus cycle was also very accurate. They observed its changing from Evening star to Morning star and it is very likely that they knew that they were the same object. After all the two were never seen together at the same time. Their positions indicated that the Morning Star "visited" with the Sun and reappeared as the Evening Star. We inherited the name of Venus from the Romans, for whom it represented their goddess of beauty, Venus.

Venus also exhibits phases, similar to those of the Moon. Those were instrumental later in discrediting the world view of all celestial objects including Venus orbiting Earth. There is a full Venus when it is on the opposite side of the Sun from Earth. Then it is of course the farthest away from Earth, hence it is small. Venus shows a Moon like crescent shape when coming around the Sun toward the Earth side. The new Venus, similarly to the new Moon, occurs when the planet is between us and the Sun. This is the time when it is the biggest, however, it is dark.

The ancient Mayan observers were cognizant of the 584 days long Venus cycle. This is the Venus year, the time Venus needs to complete a full circle and arrive at the same position with respect to the Sun, but not necessarily with respect to the sky. Of this cycle, Venus is visible for 263 days as the Evening Star and 263 days as the Morning Star. Venus is not visible for a 50 day long period in between when it is behind the Sun, and for eight days when it is in front.

This cycle is described in a rare surviving Mayan artifact, the so-called Dresden codex, so named after its final location. The codex contains ancient Mayan hieroglyphs describing a sequence of five 584 day Venus cycles. This cycle is of $5 \cdot 584 = 2920$ days, which is eight 365 day long Earth years. Hence Venus appears exactly in the same position in every eight year with respect to the stars, when viewed from the same position on Earth. The Mayans seem to have known that.

The transits of Venus, the phenomenon equivalent to the solar eclipse by the Moon, was also observed and recorded by the Mayans. Transits occur when an apparently smaller celestial body passes in front of an apparently larger object. The emphasis is on the word apparent, the actual sizes of the objects may be exactly the opposite. The opposite scenario, when an apparently larger object passes in front of an apparently smaller one, is called an occultation.

A transit of Venus occurs when it is passing directly in front of the Sun. Venus is much farther away from Earth than Moon, and the ratio calculation of its size and distance with the Sun does not produce the lucky coincidence we got for the Moon. Hence the Venus transit appears as a small dark spot moving across the face of Sun.

The intriguing aspect of the transit is its rarity. It occurs in a pattern that is repeated only in every 234 years. A 121.5 year long span is followed by two consecutive transits occurring eight years apart. Then another 105.5 year long hiatus follows. The earliest historically documented pair were in December of 1874 and in the same month of 1882. Historical records indicate that Captain Cook even observed the prior one in 1768 on his Tahiti voyage. The most recent pair occurred in 2004 and 2012.

After that humankind will have to wait until 2117, 105.5 years after 2012 to see another transit of Venus. Then eight years later, in 2125 there will be another transit, followed by the 121.5 year long hiatus. These again will occur in December as in the 1800s.

Incidentally the Venus cycle of 8 Earth years is almost exactly 99 lunar cycles, (99 * 29.5 = 2920 = 8 * 365), adding to the specialty of the relationship between the two objects, as perceived by the Mayans. The Venus cycle was very important in their lives in connection with larger than everyday events. The installation of certain rulers, religious rituals, initiation of wars were planned accordingly.

Members of the solar system were not the only celestial objects the Mayans were aware of. They clearly knew about the Star cluster Pleiades or Seven Sisters (after the Greek mythology), since they had a Mayan name for them: Tzabek. The cluster is probably the closest such object to us and it is visible by the naked eye during winter nights on the Northern Hemisphere. It is very likely that the Mayans recognized them as a group, albeit the physical phenomenon governing their togetherness was probably beyond their understanding. It is somewhat still beyond ours, although we know now that they are all moving in the same direction and with the same velocity across the sky.

The Mayans clearly have developed a science in observing the repetitive cycles in the skies and used the observed celestial cycles to measure time. There is no evidence that they recognized the rotating phenomenon behind those cycles. That recognition was due to another ancient culture, the Greeks.

3

BEING ROUND

Ancient Greek scientists like Archimedes, living on islands surrounded by the Mediterranean Sea or in coastal towns of the mainland, were very familiar with the phenomenon of ships disappearing beyond the horizon in a gradual manner. First the hull disappeared but the sails were visible for a while, then those dipped below the horizon as well. In the converse, the sails and upper structure of ships appeared first when approaching a port, finally followed by the hull.

They also observed the rising or setting of the Sun from or into the sea, depending on their location and orientation. All these observations indicated a round Earth to those who attempted to interpret these phenomena. The round nature of Earth still took a long time to be generally accepted, and there are still some people in denial of that fact.

Credit for a scientific explanation is due to Eratosthenes of Alexandria in the third century BC. He was, as scientists of his time commonly were, a polyhistor: an expert in multiple sciences. He was well versed in astronomy, mathematics, geography and even philosophy. He was the director of the library in Alexandria, by all accounts home of the greatest collection of written material at the time.

According to anecdotal evidence, he heard about the fact that in the southern city of Syene, a vertical stick placed into the ground had no shadow on a particular day at noon. The day was the summer solstice, well known to astronomers of his time. The city of Syene is the modern day Assuan that lies on the Tropic of Cancer, hence the Sun was directly overhead on that day. He discovered, however, that on the same day sticks do cast a shadow in his town, Alexandria. The only explanation he found was that Earth must be round as shown below.

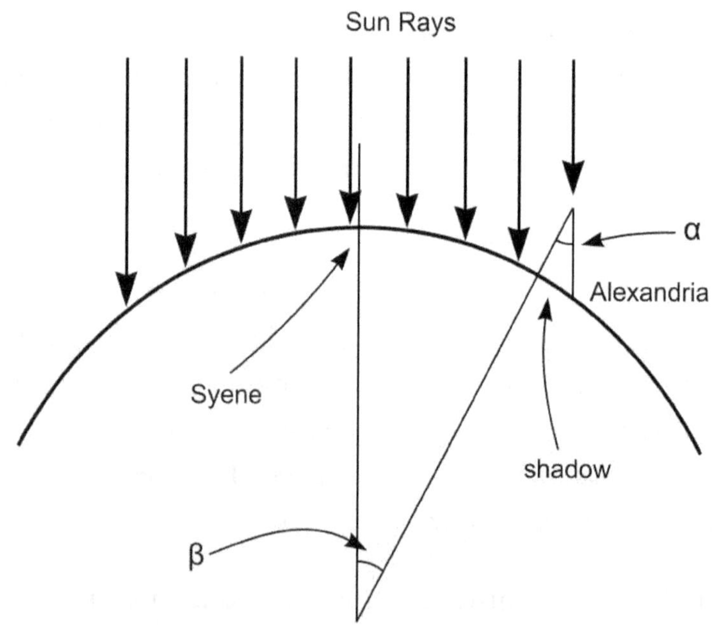

Eratosthenes' experiment

Eratosthenes even thought of an idea for an experiment to measure the circumference of the round Earth. As the story goes, he hired a person to walk and measure the distance from

Alexandria to Syene. The measurement yielded about 800 kilometers, in today's units. He measured the shadow of the stick in Alexandria and from it concluded that the angle α on the figure was roughly seven degrees, or about one fiftieth of the 360 degrees comprising a full circle.

Since angles α and β are the same by the teaching of elementary geometry, the two triangles with a curved side are similar. Therefore that ratio of the length of the shadow at Alexandria to the distance to Syene is the same as the ratio of the stick's known height to the radius of the Earth. From this he calculated the radius of the Earth to be about 6,400 kilometers (in today's units) and the total circumference of the Earth as about 50 times the distance from Alexandria to Syene, yielding about 40,000 kilometers. That is very close to the 26,000 miles we know today, which is an astonishing accomplishment of the human mind 2,300 years ago.

His results were soon widely known throughout the Mediterranean and by the beginning of the 1st century Earth was almost always depicted as a globe. This discovery enabled an entirely new interpretation of celestial observations and suddenly things were looked at in a different framework.

Incidentally, the wide acceptance and visible proof of Earth's roundness did not satisfy a group of people. They, commonly called the flat Earthers, remained in objection unbelievably all the way to this millennia. The circular shadow of the Earth on the Moon would not convince them. They would invent a circular disk description of the Earth. That would certainly explain the circular shadow, assuming that the Sun dips below the plane of the disk at night.

Their description of the arrangement of the known facts about Earth's surface was so inventive that it is worthy of a brief discussion. In its most advanced form, their belief was that the North Pole was at the center of the Earth disk and the continents were arranged in their known relations, but the perimeter of the Earth was the icy mountains of the South Pole.

One needs some imagination to follow this, but let us indulge as an exercise. The circumnavigation of Earth by Magellan in 1522, just three decades after Columbus' voyage to America, was explained by the flat Earth believers in their characteristically inventive way. In their vision it was akin to sailing around an island. They posited that the circumnavigation was only possible in East-West directions and not North-South, because then the ship would fall off at the edge of the Earth, assuming it was able to cross the icy waters and iceberg barriers. In fact the icebergs were there for the particular purpose of defending the foolish who attempted to do just that.

The strength of delusion of the flat Earthers was still highly present in 19th century England leading to one of the most famous scientific bets ever. The bet involved Alfred Russel Wallace, a self-taught British naturalist of humble origins. Wallace was born into a poor family and had only a grade school education. He spent his teenage years doing various jobs but due to science museum visits during his school years he never gave up on his dream of becoming a scientist.

Eventually he got an opportunity to be a specimen collector in the East Indies working for a botanical company. During his stay on the island of Ternate, at the time a Dutch territory, he came upon an explanation of how species evolved. He called his theory the theory of transmutation and wrote a letter explaining it to Charles Darwin, the best known naturalist in England at the time. He asked Darwin to present it to the academic society. What came next was a bit murky and there are several interpretations of it.

Darwin apparently was working on the same idea at the time, but was delaying its publication. His academic friends withheld Wallace's paper for a year until Darwin completed his own manuscript and rushed to print it. By some accounts Wallace's letter gave Darwin the missing idea he was searching for to explain his own observations. By other accounts Darwin just needed the jolt to complete his own work in order to claim priority, because he still had some unanswered questions.

Wallace, a good sport, conceded credit to Darwin, but many people now acknowledge Wallace as the co-discoverer of the evolution. Intriguingly Darwin, who went on to fame and fortune, provided financial help to Wallace later in his life, who struggled to make a living. Whether it was guilt or just friendly help, we will never know. However, Wallace's good nature also led him to his involvement with the flat Earth bet.

A Christian philosopher named John Hampden, one of the arch proponents of the flat Earth principle, originated the bet. The mainstream flat Earth believers were strongly motivated by religious beliefs and ardently adhered to the Bible. Hampden, however, having been educated at Oxford, decided to use science to prove his biblical beliefs. In January of 1870 he proposed a 500 pound bet in the Scientific Opinion, a leading forum at the time. The bet was a challenge to prove the roundness of the Earth by everyday means, understandable by a commoner untrained in the sciences.

Most of the scientific community considered the bet a joke, almost all of the people in the country ignored it, but Wallace took up the challenge. He claimed to do it as a public service to the people and in the interest of society. He may have been drawn to the amount of the prize, considering that by that time he was back in England with a family and without a stable academic job. This was not a surprise, after all he did not have a college degree and people did not yet recognize his contribution to Darwin's theory; they do so only reluctantly even today.

Wallace of course knew the phenomenon all seafarers and coastal people knew, that the surface of the ocean also curves and the ships gradually disappear under the horizon. He devised a land experiment to exploit the same phenomenon.

He chose the Old Bedford Canal, a uniquely straight man-made canal in Bedforshire, specifically a six mile long open section of it with an unobstructed view between two bridges. Wallace computed that the curvature of the Earth would be visible in a distance of six miles. He placed long poles into the water at every mile with exactly six foot long sections above the waterline at each location. The poles were also marked at every foot in height. He expected that the curvature would be shown when viewing the markings on the posts with a telescope from the bridges.

Sure enough, the experiment showed that the middle pole was about 4 feet higher than the first and the last poles at the two bridges, to which the telescope was aligned to present the horizontal "flat" Earth. The second and fourth poles also showed the proper height difference. To all rational minds and to the judge present the proof was adequate. Wallace proved the roundness of the Earth and the bet amount was released to him by the judge.

What followed was a testimonial to the irrational mind-set of the believers of the flat Earth philosophy. Hampden refused to accept the proof. In fact, he blatantly insisted that it actually proved that the Earth was flat. He continued to fight, even threatened Wallace's family. The latter act sent Hampden to jail, ultimately completely disgraced. He lived the rest of his days in poverty but never gave up his conviction of a flat Earth. Following Hampden's crusade, a flat Earth society of England was founded, later grandiosely renamed to be that of the World.

The philosophy is still with us today. The latest manifestation of the belief was the International Flat Earth Research Society of America, organized and led by a California couple. Ironically they lived in the Palmdale area close to Edwards Air Force base, the landing site of the space shuttle after completing its orbits around the round Earth. They had a regular newsletter with many subscribers as recently as the beginning of this millennia. It was a long lived faulty belief, indeed.

4

TRAVELING SPHERES

Another enduring fallacy was the view that Earth was the center of the universe, independently of its flatness or roundness. Many Greek scientists in the first millennium BC were fully convinced of the central role of Earth. The rotation of all other objects around us was duly noted, except in some cases with foolish sounding explanations.

The Aristotelian system, named after the most famous philosopher of those times, consisted of 56 transparent concentric spheres traveling around Earth. The Moon was on the innermost sphere and the reason for the large number of spheres was because they were used to justify some of the inexplicable behavior of the planets.

There were several plausible arguments supporting the view. If the Earth was moving, we should see a relative motion of the stars, which one could not. Then there was the rather obvious fact that we could not feel the air moving in our face when Earth moved. The assumed coup de grace was, however, the fact that the apparent brightness of Venus does not change and Aristotle's followers thought that could not happen otherwise but with Venus circling Earth. Of course the simple explanation of Venus as an inner planet on a concentric circle with Earth did not occur to them yet. They also glossed over the fact that Earth centered concentric spheres could not account for the changes in brightness of some of the planets that moved around in different distances from us.

One of the earliest attempts at explaining the world with the Sun at the center was made by the Greek Aristarchos in the third century B.C. He was born on the island of Samos and is still famous for his attempts at measuring the distance of the Earth and the Moon. In his book that survived the ages, titled *On the sizes and distances of the Sun and Moon*, he presented a heliocentric view. Aristarchos' measurements were a bit inaccurate, but considering the tools available for him at the time his reasoning and approach were sound.

He recognized that the Moon and the Sun have the same visual angles when viewed from the Earth during an eclipse and conjectured correctly that then their sizes should be related to their distances. He then calculated that the Sun is about seven times farther away, hence seven times bigger in diameter than the Moon. This was somewhat smaller than the now known value of about 400, but he correctly concluded that it is impossible that the much larger Sun would orbit around the much smaller Earth as fast as once a day. Therefore he arrived at his heliocentric view.

He concluded that the stars are very far away and that is the reason we do not see any relative motion between them, as opposed to the observations about Earth and our celestial partners. He also concluded that the Sun was just another star, the closest to us. He was of course on the right track, but his groundbreaking heliocentric thinking was not accepted at the time of his life.

Somewhat similarly to the religious persecution of scientists in the middle ages, he was also accused of violating the prudence of the prevailing philosophically grounded arrangement of the universe. It is unclear whether there was a formal court case against him, but he was publicly admonished by several of his contemporaries. Some simply rejected the heliocentric system because Aristarchos was not yet able to explain the peculiar motions of Venus and Mars.

The Earth centered view prevailed for centuries even after his time and was further set back by the work of Ptolemaeus. Claudius Ptolemaeus was born a Roman citizen, but of Greek or, according to some sources, maybe Egyptian ethnicity. He is now known by the English version of his name, Ptolemy, and he personally delayed humankind's understanding of our rotational surroundings for about a millennium. Despite Aristarchos' hypothesis, that was yet unproven at the time, Ptolemy convinced the world otherwise.

Ptolemy in the second century AD "proved" that the center of the universe was the Earth by explaining that the Sun, the other planets and the stars were rotating around us on concentric spheres. The spheres were actually made of crystal, as he said, something was needed to hold them up there in lieu of knowledge about gravity. Ptolemy's book titled *Almagest* became widely known and due to his academic reputation was accepted at face value for centuries.

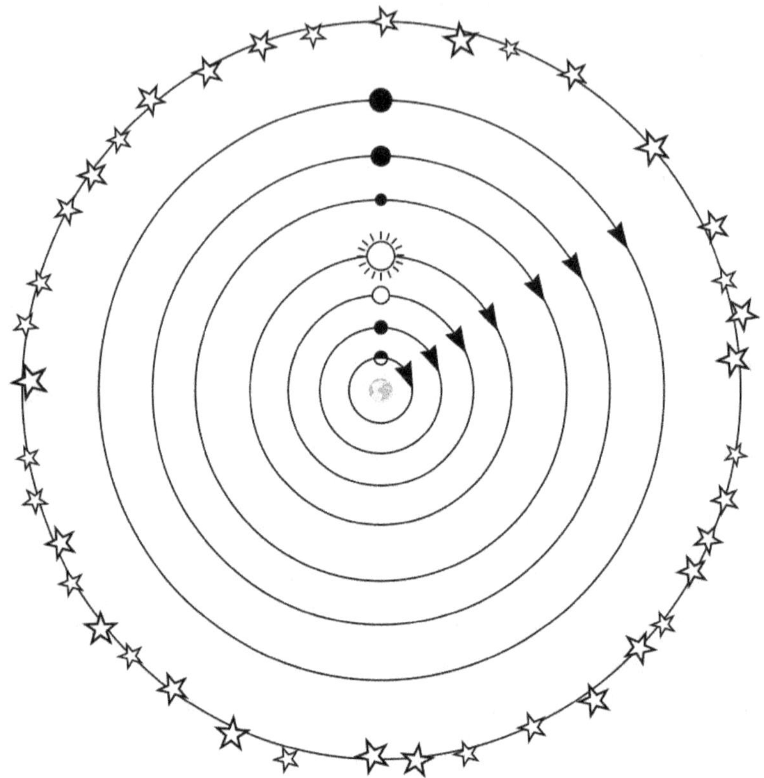

Ptolemaic universe view

In his view the order of the then known celestial objects from the central Earth outward was Moon, Mercury, Venus, Sun, Mars, Jupiter and Saturn, as shown on the figure above. They were all rotating in the same direction. It is notable that even though the Earth was supposed to be fixed in his system, the rest of the solar system was rotating. This is a testament to the early recognition of the intrinsic nature of the rotational motion.

Ptolemy of course noticed that those circular paths of the celestial objects were far from perfect. Some of them appeared to speed up or slow down, and even go backwards in certain segments. This was a serious problem of the model and the so-called retrograde motion of Mars was a thorn in his side. Finally, he invented an explanation to this by introducing another set of crystal spheres.

His idea was based on the Greek mathematicians' penchant for the circle and the wealth of intriguing curves generated by rolling circles. Rolling circles on a plane and observing the motion of certain points resulted in peculiar patterns, called cycloids. Depending on the ratio of the horizontal motion of the circle and the angular velocity of the rolling circle, an observed point located on the circle generated various curves.

The curves were smoothly waving when the speed of the linear motion of the circle was large in comparison to the angular speed. When their ratio was at a specific value equating the linear distance covered during one full revolution with the circumference, the curves created a sharp cusp. Finally, when the rotation was faster than the linear motion, the curves produced pronounced loops.

The latter shape gave an idea to Ptolemy for an ingenious explanation of the observed phenomenon. He explained that Mars is actually on a smaller crystal sphere that is traveling inside of the big sphere of the sky. This motion, illustrated on the next figure, is somewhat plausible. The partial circle with the large radius represents the circle along which the center of the smaller crystal sphere travels. The full circle represents the smaller crystal sphere. Following the point of Mars on the circumference of that circle, it will move in a looping motion of the cycloid curve shown on the figure.

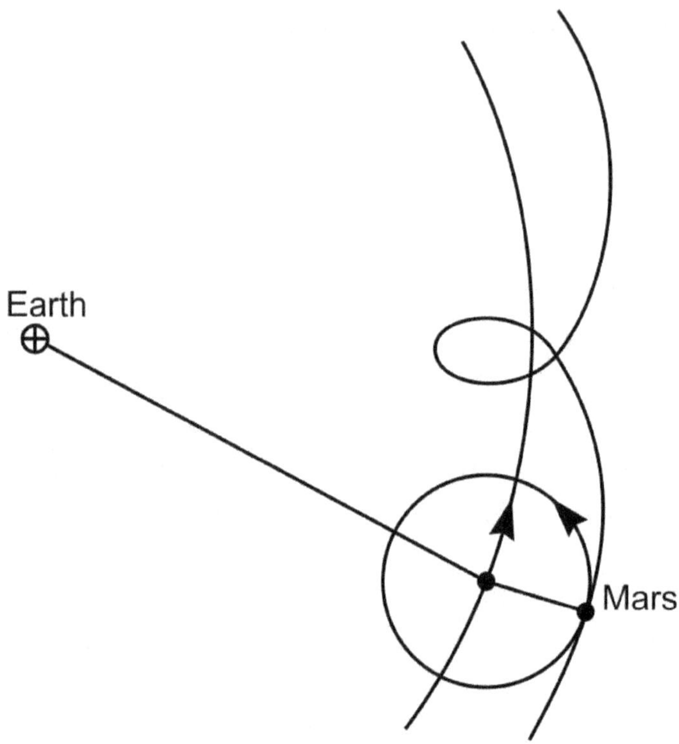

The cycloid motion

Apart from the incorrectness of this explanation, Ptolemy was an accomplished astronomer. He would observe the stars, would make notes of their locations, apparent brightness and would even predict certain eclipses. His belief that the Earth was the center of the organized rotations he observed in the sky was not unnatural and quite common in his time. It was certainly in accordance with the observations about the stability and unchanging nature of the Earth as opposed to the nightly or monthly changing heavenly phenomena.

His remaining problem was that the timing was not in compliance with practical observations. Ptolemy was never able to explain the discrepancies, nevertheless, for almost 1,500 years after his time his wrong celestial vision prevailed. Ptolemy was, however, not the only person to be faulted for this. Some of the blame is to be placed on the Catholic Church of the middle ages.

The Earth-centric vision was perfectly fitting with the biblical references to Creation and the church defended it to the bitter end. The acceptance of a moving Earth would relegate it to be just another one of the lowly planets, like Mars or Venus. This would remove it from being the center point of God's creation.

This was certainly not appealing to the Catholic church and they were staunch defenders of Ptolemy's vision until the 1500's. The second half of the 16th century became a bloody conflict over this topic, culminating in the death of Giordano Bruno at the burning stake in 1600.

Giordano Bruno was an ordained priest, hence his beliefs were considered even more blasphemous by the church. Besides his heliocentric views, he also confessed to believing in the infiniteness of the universe and the possibility of other worlds.

This was another unacceptable view to the church at the time. The scientific truth, however, could not be held back, and incidentally another member of the clergy brought the final understanding of the rotational world.

5

ROTATING WORLD

Nicolaus Copernicus was born in 1473 in the northern part of today's Poland. He put the issue of the center of the world finally to rest, although his explanation took another hundred years or so to gain full acceptance. He studied at the Cracow Collegium in the 1490s and spent the first years of the 1500s in Italy studying science and medicine. Upon returning to Poland, his uncle, the bishop of Frauenburg in the northernmost diocese of the Catholic church, took him under his wing. He appointed his nephew a canon of the diocese in 1511, a role Copernicus fulfilled until his death in 1543.

The three decades in the position allowed him to immerse into deep thinking and research about the motion of celestial objects, specifically into the discrepancies between Mars' orbit and the Ptolemy model. Copernicus discovered that by simply switching our central role with the Sun, suddenly more facts were falling into place. The observed motion of the planets would fit the model and there was no need for concocted adjustments.

It is an interesting footnote of history that the quiet cleric's work could have been lost, had it not been for the acute interest of a young German professor from the University of Wittenberg, Georg Joachim Rheticus. During the late 1530s Rheticus heard rumors about the groundbreaking work of the Polish canon and in 1539 he traveled to visit him for a few weeks. Copernicus, who never had a student as he was not affiliated with any university, welcomed the young scientist and Rheticus' weeks of visit turned into two years.

During those years Rheticus fully understood and appreciated the work, and he was able to convince Copernicus to publish it. Rheticus took the manuscript with him to Nürnberg in 1541 and arranged for printing. With the technology of the time this was extremely time consuming and took the better part of 1542. Finally, he returned to Copernicus in 1543, who was on his deathbed by that time, just waiting to see his life's work published before his death. The publication year of the book *De Revolutionibus Orbium Coelestium*, loosely translated into English as "the revolutions of celestial orbits", hence shares the year of his death.

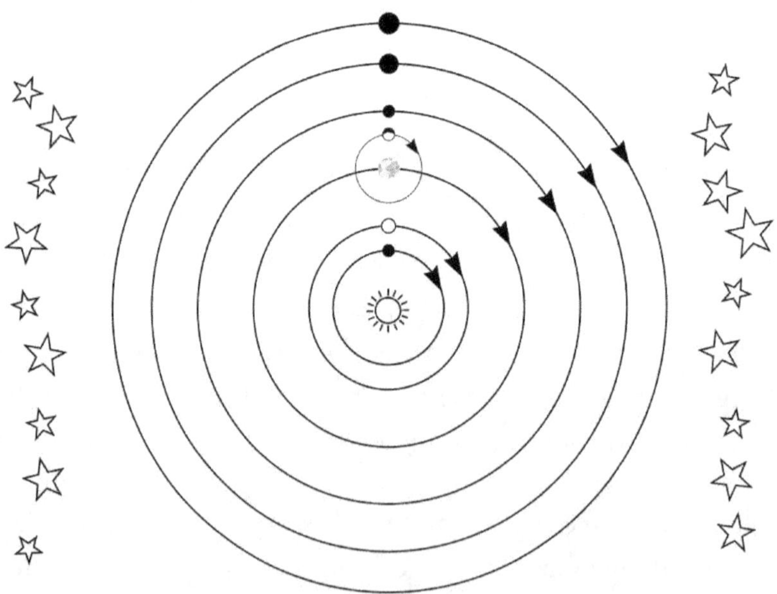

Copernicus' world view

Copernicus' world view is shown on the above figure. The order of the world now is Sun, Mercury, Venus, Earth (with Moon rotating around it), followed by the outer planets Mars, Jupiter and Saturn. The remaining planets were still not known. It is notable that the fixed stars were not anymore affixed to the outermost circle in his vision, but randomly distributed within the universe.

The view's credibility was significantly boosted by the fact that it simply explained the evening appearances of Venus. This was of course the result of Venus being an inner planet closer to the Sun than Earth. Copernicus' view also immediately resolved Mars' retrograde motion issue that required the epicyclic device of Ptolemy.

On the next figure it is clearly visible that the different relative positions of Mars and Earth produce the apparent retrograde motion. Since the Earth is closer to the Sun than Mars, counter-intuitive things happen. When the Earth is in position 1, the line of sight to Mars would show the position of Mars in point 1 in the sky.

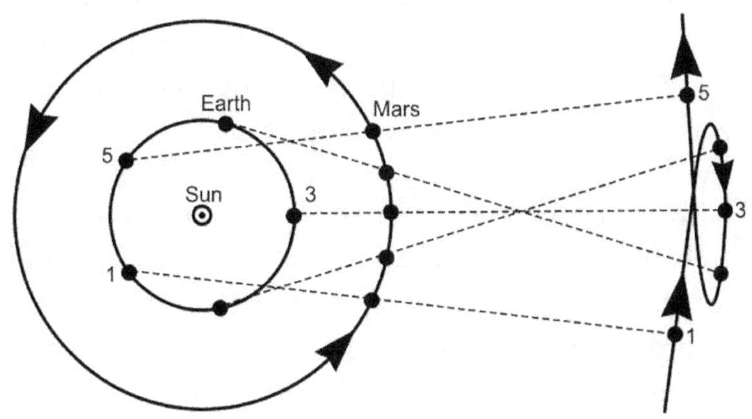

The retrograde motion of Mars

When moving through point 3, Mars is bypassed by Earth moving on an inner circle giving the appearance of Mars moving backwards against the sky. When reaching point 5, Mars has again resumed its forward motion, until another cycle is completed.

It is ironic, however, that Mars' retrograde motion was not completely resolved by the Copernican model either. Whereas Ptolemy's now obviously incorrect model placed the Mars ahead of its observed position, Copernicus' model made it delay somewhat. The acceptance of Copernicus' model now hinged on this final discrepancy.

Galileo Galilei played a leading role in the struggle to gain acceptance of the Copernican model. Galileo was born in 1564 and is often called the "father of modern astronomy". He was born, brought up and educated in Pisa, except for a few teenage years in Florence when his family lived there. He undertook medical studies at the university in Pisa obeying his father's request, but studied mathematics for his own interest. Ultimately he even received a professorship of mathematics at the university. After only three years he moved to the university of Padua for two decades of fruitful work during which he made many of his now famous observations.

Galileo, who became the leading astronomer at the time and probably the most influential of all times, started exploring the sky with a telescope. According to anecdotal evidence, he only heard about the concept of the telescope and based on that he built his own. His telescope produced a twenty fold magnification and opened up his celestial horizon to include objects that had not been seen until then. His telescope would also allow him to narrow the aperture and observe very bright objects with more definition. Galileo may have been the first person to say: "Seeing is believing".

This led him to observations of Jupiter during the early part of 1610 and soon he discovered that there were four moons rotating around Jupiter. He named the moons after the Medicis, the noble family sponsoring his scientific pursuits. His observations also proved that our Moon does rotate around the Earth, even if the Earth is rotating around the Sun. Galileo published his findings in 1610 in a book titled *Sidereus nuncius*, translated as Starry Messenger, and he became an overnight scientific sensation, even with the limited speed of information travel of his time.

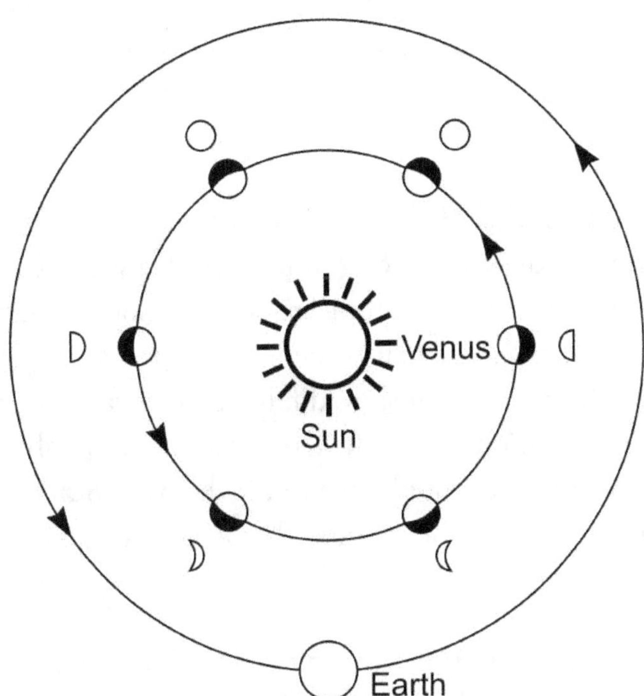

The phases of Venus

He continued his celestial observations for the next two decades. He also observed the phases of Venus, shown on the figure above, to be in agreement with a Sun-centric arrangement. Galileo explained that if Venus were on a circular orbit around Earth the sequence of the phases of the Venus could never match the observations. Since Venus could have not been rotating around Earth, hence the phases of Venus contradicted the Ptolemaic vision.

He considered this fact to be the final proof of the Copernican view. He published his thinking in 1632 in a book titled *Dialogue concerning the two chief world systems* and that finally drew the full wrath of the Catholic church on him. He was also charged with heresy like Giordano Bruno, but due to his personal connection to the Pope his life was spared. The Pope would commute Galileo's sentence on the condition that he recanted his views and confined him to house arrest for the last decade of his life.

The reason that the idea of Earth rotating around the Sun and its own axis took so long to be accepted was, however, not only due to the religious objection. Intuition was against it. People thought that when someone dropped a weight from a tower on the west side, it should fall farther away from the wall, due to the rotation of Earth. Conversely, the weight dropped on the eastern side would have to hit the wall. Since neither of these happened in experiments, the belief was held that Earth does not move.

Galileo died in 1642 and by the end of the 17th century the rotating Earth belief was widely accepted. The Vatican, however, took another 300 years to make an official statement about the church's mistakes regarding Bruno and Galileo. It took the arrival of the non-Italian Pope, John-Paul II to acknowledge "the error of the theologians of the time". In his statement he singled out the treatment of Galileo as one of the mistakes committed by the zeal of the inquisitors.

The still remaining discrepancy of Mars' motion, its apparent delay from the model was not due to Copernicus' Sun-centric model. It was due to the fact that the orbits in question were not circles, a fact only recognized about a century later.

6

ELLIPTIC ORBITS

Johannes Kepler was born in 1571 in Germany and was educated in the protestant school of the town of Maulbronn. He later studied at the university of Tübingen with the goal of becoming a priest. He was, however, more interested in astronomy and was introduced to the then radical ideas of Copernicus. Before ordainment, he took on a job opportunity as a high school mathematics teacher in Graz, Austria, hence he is sometimes referred to as Austrian.

His double interest in astronomy and mathematics made him recognize the coincidence that there were five other planets besides Earth and there were also five regular polyhedra. The five extraterrestrial planets known at his time were Mercury, Venus, Mars, Jupiter and Saturn. The regular polyhedra have regular polygon sides, like the hexahedron or the common cube with square sides. Others are the tetrahedron, the octahedron and the icosahedron with four, eight and twenty equilateral triangle sides, respectively. The fifth one is the dodecahedron with twelve pentagonal faces. Kepler imagined an intrinsic connection between the two facts and created a vision of the universe where the known planets were all nested in one of the regular solids.

We now know of course that this was incorrect, however, his work raised enough attention in the scientific circles that in 1599 he was invited by the famous Danish astronomer Tycho Brahe to Prague where the latter was the court astronomer of German Emperor Rudolf. Brahe had amassed a wealth of celestial observations during his illustrious career, but he was somewhat of a rowdy personality. He was involved in a duel and almost died of a sword inflicted wound. He also had his own hypothesis about the order in our solar system.

This otherwise head-strong and aggressive person in this particular case settled to making a compromise between Ptolemy and Copernicus. Or maybe he was astutely judging the society and erred on the side of political correctness. Brahe retained the Earth as the center of universe and kept the Sun orbiting around it, thereby appeasing the religious authorities. Quite cleverly, however, he placed the other planets onto an orbit around the Sun. This allowed a better agreement with his precise observations without making the clergy irate.

Brahe was rather disorganized and he invited Kepler to assist him in organizing his observations. Two years after Kepler's arrival, however, in 1601 Brahe suddenly died of an over-indulgence in eating and drinking (another apparently age old trait of humankind) and Kepler was named royal astronomer. He inherited Brahe's life time observation collection and that was the basis of his work about Mars.

Brahe's celestial model did not solve the Mars problem of Copernicus' model either, just as Copernicus did not fully solve Ptolemy's Mars problem, but his observations provided Kepler the data to prevail. He worked on this topic for the better part of a decade before he realized that the common assumption made by all three of the universal models was wrong. They all assumed that the planets rotate in circles with constant speeds and with either the Earth (in Ptolemy's vision) or the Sun (in Copernicus and Brahe's visions) in the center of the rotation. They were all wrong!

Kepler realized that planets move along ellipses, not circles, with the Sun as one of the focal points of the ellipse. Ellipses as geometric objects were well known by the ancient Greek geometers as a member of the class of curves called conic sections. These curves are obtained by cutting a circular cone (like an ice cream cone) by a plane in various directions. Cutting a circular cone perpendicularly to the axis produces the familiar circle, while cutting it somewhat slanted results in the elongated circular shape of the ellipse.

Kepler's ellipses were still in line with the Copernican view of the planets rotating around the Sun, after all circles were just special cases of ellipses. It also helped that the ellipses in question were very much circle like, at least in the case of the inner planets. This was an extremely important distinction, however, especially so in connection with Mars.

Mars' orbit that caused so much grief for a millennium, troubling both Ptolemy and Copernicus, had a ratio of the two axes as 0.9956. This is very close to one so it is almost a circle, but just elliptic enough to make the earlier models inaccurate. Finally, the issue of Mars' motion was correctly explained in his book titled *De Stella Martis*.

The remaining issue bothering Kepler was the apparently uneven cycles of various planets during the duration of one turn. He asked: Why would they accelerate and decelerate?

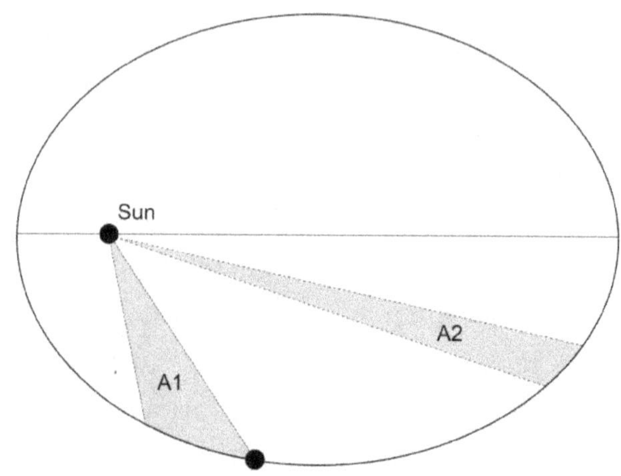

Kepler's ellipse

Kepler concluded based on his calculations that the areal velocity of the planets is constant in rotation. That means that the sections denoted by A1 and A2 on the figure above have equal area. When the planet, depicted by the dot on the circumference of the ellipse, was in close proximity to the focus the Sun occupied, it was moving faster and covered a wider swath of the orbit. Due to the proximity, however, the lines between the Sun and the planet position were shorter.

On the other hand, when the planet was on the opposite side from the Sun, the lengths of the lines were much longer, hence the angle of the section narrower. The two areas, however, were the same. Therefore the orbital speed of the planet was smaller on the far side. This recognition was quite extraordinary, based on pure observations, attesting to Kepler's genius. This became Kepler's second law.

Kepler also explained how several planets could occupy distinct elliptical orbits around the same Sun. He realized that they are located in a slightly different orbital plane of the solar system, hence they can peacefully coexist in the glare of the common Sun.

Kepler published his new world order in a book titled *Astronomia Nova* at the end of the first decade of the 17th century. Not surprisingly the leading scientists of the time did not endorse it. After all, circles were so beautifully symmetric, why would God create some ugly squashed circles? Some even attempted to re-create the elliptic pattern, that was now indisputable from Brahe's observations, with circles and cycloids like Ptolemy. The world had to wait for Newton to prove that Kepler was right.

Kepler also observed that the outer planets, farther away from the Sun orbit on less circle like, elongated ellipses and at higher speeds as well. This observation led him to his third law stating that the square of the orbital period of a planet is equal to a constant times the cube of the mean distance of the planet from the Sun. The mean distance is the average of the distance between the planet's closest (perihelion) position to the Sun and its farthest (aphelion).

According to this law the square of the orbital period of a planet around the Sun is $P^2 = K\,P^3$, where D is the mean distance and K is the so-called Kepler constant that depends on the mass of the planet. The value of K for Earth is $4.03 \cdot 10^{-29}$. Earth's mean distance from the Sun is about 150,000 kilometers, or $1.5 \cdot 10^{11}$ meters. Using the formula above results in our well known orbital period of 365 days.

Planet	Distance	Period
Mercury	0.387	0.241
Venus	0.723	0.616
Earth	1	1
Mars	1.524	1.88
Jupiter	5.203	11.9
Saturn	9.539	29.5
Uranus	19.191	84.0
Neptune	30.071	165.0
Pluto	39.457	248.0

The table shows the mean distance from the Sun and the orbital time computed by Kepler's third law for all the planets now known in our solar system. The numbers are normalized relative to the Earth's motion. The period is measured in years and the distances are in astronomical units, the mean distance of the Earth. Ellipses of our solar system are truly enormous!

Kepler's law for Mars with $K = 3.95 \cdot 10^{-29}$ and using Mars' mean distance from the Sun as 225,000 kilometers yields an orbital period of 684 days. The next planet in our solar system Jupiter has a Kepler constant of $K = 3.98 \cdot 10^{-29}$ but with over five times bigger mean distance. Jupiter will observe a rather long year with 4,331 days.

The distances and times of the outer planets are mind-boggling. For example, for Neptune the table indicates a mean distance of about 30 times that of the Earth and a 165 Earth years long orbital time. Poor Pluto is even farther out and recently had to suffer the disgrace of being removed from the list of planets by some newly established planetary standards.

The most radical demonstrations of the elongated elliptical orbits are the orbits of comets. This was recognized by an English astronomer, Edmund Halley, about a hundred years after Kepler. He was born in 1656 to a wealthy family enabling him to pursue scientific interests, specifically astronomy. He knew of medieval written records about a bright, fast object occasionally appearing in our solar system at a very steep angle compared to the orbit of the planets and disappearing again forever.

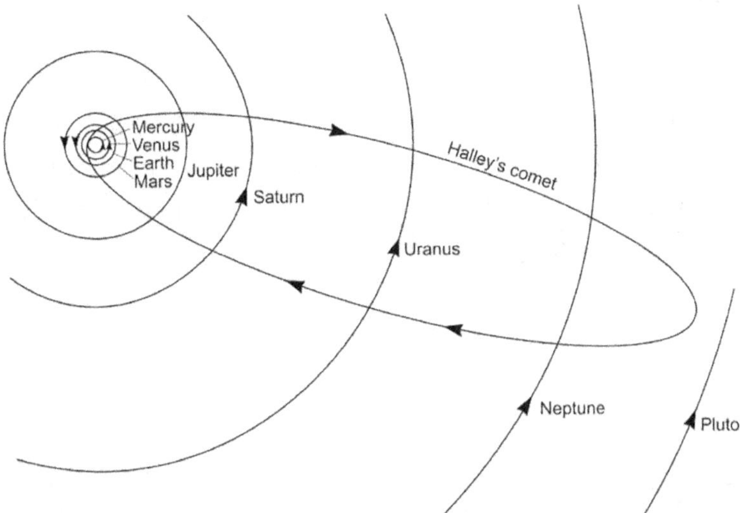

Halley's comet

Halley recognized that those records may be talking about one particular object regularly appearing, but with a very long orbit time; a comet. The comet took an appearance during Halley's life in 1682 and Halley observed its path carefully. He predicted that the comet would reappear in 1758, when he would be 102. Such prediction was not really taken seriously, until the comet did reappear as Halley predicted. The object was then named Halley's comet.

The comet whose rotational path is shown on the prior figure, takes about 76 years to complete its elongated elliptical orbit. When it is the farthest away from the Sun, in its aphelion position, it is about 5.3 billion kilometers away, almost reaching Pluto's orbit. In its perihelion position it is inside even Venus' orbit only about 90 million kilometers from the Sun. This is an extremely long axis of an elliptical orbit, essentially the radius of the solar system, especially now that Pluto has been pulled from the ranks of planets.

Even Kepler's genius in observations and mathematical explanations did not extend to the understanding of the cause of the motions. He held some vague belief about a "magnetic" connection between them. The understanding ultimately came from Halley's friend, Isaac Newton.

7

ROLLING BALLS

The first scientific explanation of the falling phenomenon is credited to the Greek philosopher Aristotle. Aristotle was born in 384 BC near Thessaloniki, already a city then and still striving today. He went to Athens to study in Plato's Academy when he turned 18 and spent another 18 years of his life under Plato's tutelage.

After that he was appointed to be the teacher of the son of the Macedonian King, Alexander (the future Great). After teaching and advising the young Alexander for 8 years, he returned to Athens and founded his own academy, called Lyceum. He wrote most of his scientific work in this school during the next dozen years of his life. One of his publications was titled "Physics" and this is where the subject of the falling phenomenon was first discussed in detail.

Aristotle recognized that the medium in which the object is falling influences the speed of the event. He noticed the difference between the speed of a rock falling in air as opposed to in water. This is in a sense still valid, although as we know now, the difference is between the resistance exerted by the media, not in the difference in the phenomenon.

Aristotle also correctly observed that a falling object increases its velocity during the fall. He just incorrectly concluded that an object twice as heavy as another one will fall twice as fast. This was mainly due to the fact that the falling phenomenon was fast and the differences were immeasurable by the time measuring devices of the time. It took Galileo's genius to devise a way to slow down the phenomenon to be able to measure it as we will see it now.

Galileo, besides his already mentioned celestial observations, was also interested in the falling phenomenon. To be able to measure the process, he created an experiment with a ball rolling down on a slightly inclined wooden board under the force of gravity alone. We don't know the exact dimensions of his board, but the inclination was relatively small compared to the length of it as shown on the figure.

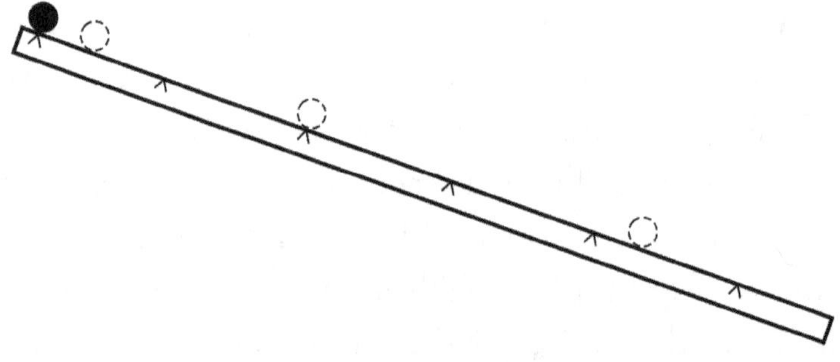

Galileo's experiment

The pace of the ball rolling down on the slope was measured by a time keeping device. It could have been a water clock, a practical device to measure the speedy event of the experiment. The water clock was a vessel filled with water and had a hole through which the water drained. The level of water in the vessel, usually observed with the aid of a floating marker, was indicative of the time passed. The resolution of the clock was controlled by the size of the hole that could be adjusted.

It could also have been a pendulum type device, as Galileo was the first to propose such and will be discussed in a later chapter. The device needed a fast pace, probably about a second between time periods, enabling Galileo to follow the path of the rolling ball.

The location of the ball was marked with a chalk at the end of each time period. After completing the experiment, Galileo compared the distances between the marks. He found that the distances covered by the ball at the end of the second period was 4 times the distance covered in the first period. The distance in the third period was 9 times that of the first and in the fourth period 16 times longer.

He concluded that the motion toward Earth was increasing its distance proportionally to the square of the time, since the distances were $4 = 2^2$, $9 = 3^2$, $16 = 4^2$ units and so on. It is known from his writings that he did not use the modern word gravity, he used something that loosely translates into heaviness. He wrote about the heaviness of objects, but from now on we will be using the modern terminology.

Galileo executed rolling experiments with balls made of different materials and the results did not change. He conjectured therefore that the rolling balls have the same constant acceleration produced by gravity. Galileo also understood that the time required for a body to reach a certain velocity from stationary position with a constant acceleration is the same as the body would need to travel the same distance with half of the final velocity.

Combining the two observations, he arrived at the value of the acceleration of gravity. Galileo succeeded to mathematically describe the phenomenon he measured. Galileo did not use the g notation but we will use this as it is the current standard. There are also some who argue that the experiment could not have produced accurate enough results due to lack of accurate time keeping devices of his time. Nevertheless, some records indicate that he was very close to 10 m/sec^2. We now know that g is 9.81 m/sec^2 in metric and 32.2 ft/sec^2 in English units.

Whether his rolling experiments enabled him to calculate an accurate value of the acceleration is immaterial. The fact that a constant value was at work was the important recognition. Galileo considered the constant to be the same anywhere on the surface of Earth, but we will see in the following chapters that there are some variations, some of them rather intriguing.

Then Galileo concluded that when the cause of the acceleration is exclusively gravity, like in the case of falling bodies, the loss of height of the falling body is computed by the formula: $h = 0.5 \cdot g \cdot t^2$, where h is the distance falling in time t. This formula is now known in even first grade high school physics as the equation of free fall.

According to a sometimes disputed story, Galileo in the early 1600's also executed an experiment at the leaning tower of Pisa. He demonstrated that a wooden ball with the same shape as an iron ball reaches the ground at the same time, when dropped from the high tower. This proved that the acceleration of a body in free fall is constant, independent of its weight, assuming the shape of the body does not generate a measurable amount of air resistance.

This story may be false and the event may have never happened. Clearly it was not the experiment that led him to his equation. It may have been though actually executed to demonstrate his findings to a lay audience later. Or, it may have been a duplication of an experiment he heard about. Simon Stevin, a Dutch scientist allegedly executed the same experiment at the church tower of Delft, some twenty years earlier. Galileo was not opposed to follow other people's ideas, after all it is well known that he built his first telescope after hearing about such built by another scientist, incidentally also a Dutchman. Whichever was the real cause we will never know, one thing is certain; the Pisa tower experiment of Galileo is now forever embedded in the history of science.

Galileo was also interested in the motion of projectiles, a topic of interest later when we consider shooting rockets. He decomposed the motion into horizontal and vertical components. He assumed that the distance traveled horizontally is simply proportional to the velocity. By that time this distance versus velocity relationship was well known, since even everyday persons knew that a faster horse travels a longer distance during a day than a slower animal.

Galileo further understood the concept of inertia, even if he did not call it such. He knew that a body would continue with a constant speed until its motion is modified by a force. If an object is thrown horizontally in an environment without any friction or resistance, it will move with a constant velocity indefinitely.

On the other hand the object will undergo the same motion vertically as if it were just dropped and he was the one who derived the formula for that. Galileo recognized the fact that the time is common between the horizontal and vertical motions. By equating the time of the two motions he proved that the shape of the trajectory is the well known parabola. That meant that the height change of the projectile was also proportional to the square of the horizontal distance, assuming that the effect of the air resistance is negligible.

Galileo therefore could compute the maximum height and the maximum distance attained by any projectile. His knowledge soon found its way into military science and became the foundation of ballistics. It was applied to calculate trajectories of cannon balls and other projectiles of warfare.

Examining the case when the projectile is shot directly upwards, he realized that the motion due to the initial velocity is countered by gravity. Once the velocity produced by the acceleration of gravity is the same as the initial velocity, the object will reach the top of its trajectory. It will then fall back again and will hit Earth with velocity equal to the initial velocity. Hence Galileo recognized the potential energy manifested in the elevation of an object. He was a step away from establishing the concept of the gravity force, a topic of much importance later.

8

FALLING MOON

Isaac Newton was born on a farm in Lincolnshire, England in the year of 1642. He was a prodigious student and enrolled at Cambridge University soon after finishing his grade school studies. In the late 1660s, during a cholera epidemic in London, he returned to the family farm for about a year. According to a now enshrined anecdote, possibly as fictitious as Galileo's Pisa story, one evening having tea in the apple orchard behind the family house he noticed an apple dropping and hitting the ground.

Newton realized that a force of Earth was pulling the apple to the ground and started to contemplate how far the force reaches above the apple tree. He concluded that it is likely to reach the Moon. He extended this thinking and considered that even Earth itself is pulled by a similar force, that of Sun's. He wanted to quantify these relationships and needed to find out how this attraction force changed with distance.

According to historical records, the idea of the force of gravity changing by the square of the distance was originally proposed by Robert Hooke, the inventor of the law of springs. Hooke was, however, an experimental scientist without the mathematical acumen of Newton who ultimately got the credit for it by actually proving the correctness of the conjecture. This resulted in a lifetime of animosity between the two of them, but not the only one of such in Newton's life. He was apparently rather selfish and ruthless in such circumstances, but we will leave that topic for the biographies. His brilliant insights into Moon's rotation is our subject.

The way Newton came about proving the conjecture is worthy of discussion. The radius of Earth and the orbit of the Moon were well established by Newton's time and from this he calculated the length of the orbit of the Moon as the circumference of a circle. The time it takes the Moon to do one rotation, the length of a lunar month, was also well known. He then computed the orbital velocity of Moon by dividing the circumference of Moon's circle with the time Moon takes to complete it.

On the other hand Newton concluded that if on Earth an apple falls about 16.1 feet per second (since the acceleration of gravity is 32.2 and the free fall formula of Galileo uses half of that) then the apple (or any other object for that matter) falling from Moon's orbit (that is about 60 times farther than the radius of Earth) would fall 1/3600th (3600 being 60^2) of 16.1 feet in one second. This is equally true for the apple or the Moon.

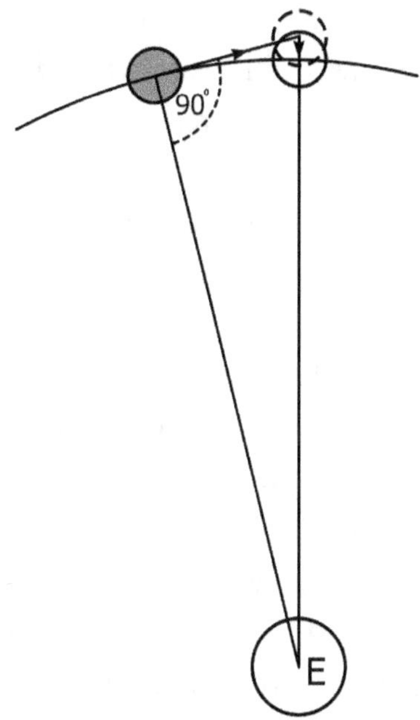

The falling Moon

Newton sketched the scenario shown on the figure above. The arrow tangential to the orbit represents the direction Moon would travel without gravity due to its orbital velocity. The length of the arrow is the distance Moon would travel in one second. The smaller arrow pointing toward Earth is the amount Moon would fall due to gravity.

He understood that if Earth's gravitational pull would be removed, Moon would continue uninterrupted along a tangential direction to its orbit. On the contrary, if Moon became stationary by losing its orbital velocity, it would fall back to Earth. The continuing pull of Earth keeps Moon in orbit by its direction pointing always toward Earth independently of the position of Moon on its orbit.

Using the Pythagorean theorem in the right triangle of the figure Newton was able to verify that the result was astonishingly close to the expected value. Newton therefore proved in connection with Earth and Moon that gravity is diminishing by the square of the distance; a spectacular feat of logical thinking.

Newton went back to Cambridge and spent years refining the theory based on his lunar calculations. His final law of gravitation states that two masses will pull on each other with the force that is decreasing proportionally to the square of the distance between them. His law of gravitation states that two bodies, with masses M and m and a distance R between them, will pull on each other with the force of $G \cdot M \cdot m / R^2$, where G is the universal gravity constant, a difficult to measure tiny quantity of $6.6 \cdot 10^{-11}$ in the metric system of measurements where the masses are in kilograms and the distance is in meters.

More interesting about this constant is that it is intrinsic to our universe and applicable between any pair of masses. Imagining M being the mass of the Earth and m the mass of the Moon, the force describes the falling Moon phenomenon. Interestingly the formula says nothing about the volume or shape of the objects, an appealing but tough to accept fact.

Newton explained that the mass of an object defines its inertia, a topic of some intrigue in later chapters. If an object had twice the mass of another one, it also resisted a motion two times more strongly. This observation led to Newton's second law of motion stating that the force required to move an object is proportional to its mass and acceleration, the now famous $F = m \cdot a$.

Newton also conjectured that the force is acting across astronomical distances. It has the longest reach but the weakest magnitude. It is an exclusively attractive force that pulls any two objects with masses together.

Therefore each object in the solar system has its own gravitational field which influences the other objects in its vicinity. The force of course diminishes by the square of the distance between the objects' center of mass, and the larger the masses are the bigger the force is. But his main and most influential realization was that the planets are kept in rotation by the force of gravity!

Newton's universal gravity law applied to a body on the surface of Earth simplifies, since the distance R becomes the earth's known radius and one of the masses (M) is that of Earth. Introducing the term of the acceleration of gravity as $g = G \cdot M / R^2$, where the gravitational pull of the Earth on a body, also known as its weight, becomes the well known formula of $W = m \cdot g$. The form is now in accordance with Newton's second law of motion stated above.

Hence the acceleration of gravity on any object on the surface of Earth becomes the value g that Galileo measured. The gravitational force of Earth on a body then becomes its weight. Since different bodies have different masses, their weight will also differ. Seems like Newton and Galileo were in perfect synchrony.

The assumption of Newton's theory that all the mass is concentrated at the center of Earth is of course physically not true. He also recognized the fact that on the surface of other planets this value would be different due to the different mass of those planets. We now know for example that on the Moon, the acceleration of gravity is about 1/6th of that of the Earth, or 1.63 m/sec^2, while Sun's acceleration of gravity on its surface is 274.1 m/sec^2 or almost 30 times that of the Earth.

The application of Newton's law of gravity to the solar system enables us to compute the masses of our planetary neighbors based on their visible rotational orbits and measurable orbital times. This branch of science is called gravitational astronomy and it is used to compute masses of celestial objects with surprisingly high accuracy without having ever traveled there, not that being on the surface of the planet would provide any better means to measure its mass.

In space with no air resistance a planet will retain its orbit due to the force of gravity acting on it. In fact the law describing the orbit of a planet around the Sun, Kepler's first law, may be derived mathematically from Newton's law of gravitation, but it exceeds the intended level of mathematics in this book. The ability to derive Kepler's equations from Newton's gravity formula was considered to be the ultimate proof of the correctness of both theories.

In the case of a single planet and the Sun, the orbit derived is an ellipse. The period of the orbit depends on the size of the orbit. However, in our solar system there are more planets. The influence of the planets on each other, besides their Sun controlled motion resulted in some aberrations from the single planet orbits leading to the discovery of some of the yet invisible objects of our rotational world.

Such unexplained motion of Uranus, for example, was attributed to the presence of another celestial body outside of its orbit. Newton's theory of gravity was able to predict the existence of a planet not known yet. The French scientist Urbain Le Verrier described the anomalous motion of Uranus with a system of 279 equations and he even solved it manually. Sure enough, the German astronomer Johann Galle in the Berlin observatory, to whom Le Verrier wrote about his solution, discovered the planet Neptune almost exactly in an orbit where Le Verrier's computations indicated it should be.

The rotational orbit of a planet would be completely unchanging if it was the only planet around Sun. That would place the position of the planet closest to the Sun, the perihelion, fixed. Due to the effects of the other planets in the solar system, this position shifts over time. This is rather normal for all planets in the solar system, Earth also has its own perihelion shift resulting in its approximately 25,760 years long celestial cycle.

Such a perihelion shift, also called perihelion precession, led to the questioning of the correctness of Newton's theory. The shift of Mercury's perihelion by Newton's law, computed in the 1860s by Le Verrier was as angle of 531 arc seconds (one 3600th of a degree). The observed shift was about 574 arc seconds a century, a difference of 43 arc seconds.

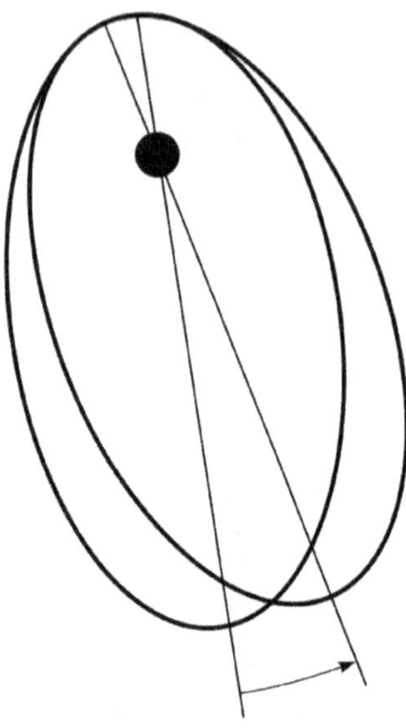

The perihelion precession

The perihelion precession of Mercury is indicated by the arrow in the next figure, where the dot is the Sun and the ellipses are Mercury's orbits. Assuming the absolute validity of Newton's law of gravity, Le Verrier was on the hunt again. He was so much convinced that the law was inviolable and there must be a planet inside Mercury's orbit that he even named it in advance to Vulcan. Despite some false findings, however, the planet was never found. This opened up the room for another theory of gravitation leading to Einstein.

Einstein's theory about two hundred years later was able to account for the 43 arc seconds discrepancy, hence it was considered to be the new "rule of the universe". But Einstein's theory now also has difficulties accounting for some observed phenomena, related to the Pioneer probes we sent into space that do not seem to observe Einsteins laws either. Hence there are now new efforts to rehabilitate Newton's law.

One of the possibilities is to allow the gravitational constant of Newton, *G*, to be varied. It could vary in distances very close to or very far from large gravitational masses, after all, we only have one place to measure it in the solar system. If it would be somewhat different in the close proximity of the Sun it could even explain the Mercury anomaly. Due to the scientific entrenchment of Einstein's theory, such efforts are highly disputed and we will not follow them.

There are also some voices in the scientific community stating that Einstein's calculations of the Mercury precession are actually false. They propose that the calculations of Einstein were actually based on incorrect observations of the rather infrequent phenomenon. Mercury's transit occurs sporadically in a pair of events separated by 3 years. Obviously Einstein's comparison calculations were also based on the 3 year pairs.

It is, however, now recognized that one of the pair is always in May and the other is in November, and that poses a problem. In May Earth is below the ecliptic and in November it is above, and since the observation of Mercury's perihelion is from Earth, the three year pairs are actually not comparable. The paths of Mercury leading up to and following the transit have distinctly different slopes, one descending and the other one ascending. It is possible that Einstein proved an incorrect observation, hence his theory does not really provide the exact result.

This topic is still undecided, since it boils down to finding the ultimate reference system in which the exact observation and calculation should be made, and that is not easy. It appears that if one applies Newton's theory in a rotating reference frame including centrifugal and Coriolis forces, Einstein's result can also be achieved.

Newton was deeply disappointed that he was not able to explain what carried the force of gravity, but he was able to explain another interesting repetitive Earthly phenomenon: the tides. In his work titled Principia he proposed that the gravitational forces exerted on Earth by Moon and the Sun are responsible for the phenomenon.

9

SPINNING EARTH

The spinning of Earth around its own axis was hypothesized by ancient astronomers for millennia. Proving it was, however, not an easy feat. The biggest problem was that one could not feel the rotation of Earth and it is not a surprise; we cannot feel it today either. How can one prove the rotation of Earth? It was a question that baffled scientists for centuries until finally in 1851 the French Foucault succeeded in answering it.

Léon Foucault was born in Paris in 1819 and he was another scientist who started studying medicine and turned to physics later. In his case it was mainly due to his late recognized fear of blood, certainly a hindrance when one wants to pursue the medical profession. His first physics interest was the speed of light and he attempted to measure it. His result was surprisingly good at 298,000 kilometers per second, considering the time and the equipment available to him.

His experiment in 1851, demonstrating the rotation of Earth around its own axis, brought him instant fame and plenty of accolades. He received the medal of the Royal Society and was named the Physicist of the Royal Observatory in Paris. Foucault used the dome of the Pantheon building in Paris to suspend a

28 kg weight with a 67 meter long wire. The long pendulum was released from one side of the building and, as was known from prior experiments, it was expected to hold its plane of swinging. The pendulum, however, apparently continuously changed its plane, demonstrated by its tip drawing a pattern in a sand pit.

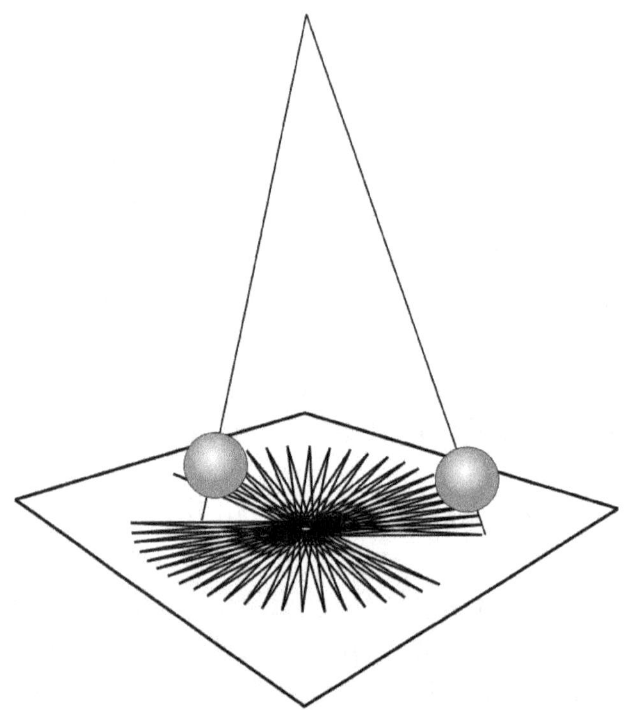

Foucault's experiment

The plane of the pendulum's swing, as shown on the figure above, rotated clock-wise about 11 degrees per hour, hence it was highly visible with a small duration of observation. The pendulum completed a full circle in about 32 hours. Since the experiment is conceptually simple, it provides the basis of many modern day Foucault pendulums at observatories world-wide. Two prominent examples are the one at the Smithsonian

Museum in Washington, D.C. and another one at the Griffith Observatory in Los Angeles. Most modern versions knock over dominoes standing on the perimeter of a circle underneath a dome. Having a domino at every 5 degrees, the pendulum will knock one over roughly in every half an hour to the delight of the visiting audience.

The actual duration of the experiment depends on the latitude of the location where it is executed. At the North Pole, for example, the complete circle would take exactly 24 hours. Conversely, a pendulum suspended on the equator would not show the phenomenon at all. In both cases the plane of the pendulum swing is retained of course, but Earth's rotation produces the different outcome.

At other latitudes the pendulum's swing is somewhere between the polar and equatorial behaviors, and the time of completing the circle varies. The change in the angle of the plane of the pendulum at a latitude angle of φ is proportional to $sin(\varphi)$. Specifically, the change in the rotation angle of the plane is $360 \cdot sin(\varphi)$ degrees per day.

At the pole the latitude angle is 90 degrees, where the sin function is one. Hence the full rotation of 360 degrees takes a day. At the equator the $sin(\varphi)$ is zero and there is no change of the plane of the swing. At an intermediate latitude the $sin(\varphi)$ is less than one, resulting in less than 360 degree rotation a day. Consequently, more than 24 hours are needed to complete the circle, like the 32 hours at the Paris latitude.

Having proven the rotation of the Earth around its own axis brings us to the question: How does Earth maintain its rotation after millions of years? Why are the orbital rotations of Earth around the Sun or the Sun along with our planets around the center of our galaxy still continuing and not seeming to slow down?

Well, the answer lies in this magical rotational quantity called angular momentum and the principle of its conservation. The angular momentum of a particle with mass m and rotating with a circular velocity v at a distance r from the axis of rotation is $m \cdot r \cdot v$. The circular velocity of the spinning particle is well known from high school physics as $v = r \cdot \omega$, where ω is the angular velocity. With this the angular momentum of the particle becomes $m \cdot r^2 \cdot \omega$.

The conservation principle is demonstrated in everyday circumstances by the spinning body of an ice skater. When extending the arms away from the body the rate of the spin slows down and when the arms are pulled close to the body the spinning accelerates. This is because the angular momentum contained in the spinning body remains constant, ignoring the small friction arising in the connection with the ice. Extended arms mean larger radius of the body and conversely, pulled in arms a smaller radius. Since the mass of the skater does not change, the angular velocity must, accordingly.

For the Earth we need to take into consideration all particles and their locations. This requires a bit more algebra as there are a considerable number of particles in Earth and their location is not always on the surface. Hence, the integration of the above formula throughout the volume of the Earth results in approximately $M \cdot R^2 \cdot \omega$, where M is the total mass of the Earth and R is its mean radius. If we use Earth's radius as approximately 6 million meters, the middle term in our equation becomes roughly $36 \cdot 10$ to the power of 12, or thirty six trillion.

We can compute Earth's own angular velocity from the fact that it takes 24 hours to make a full revolution of 360 degrees, or 2 pi radians, hence it is about $\omega = 2\pi / (24 \cdot 3600) = 0.000073$ radians per second. A radian is the angle one obtains in a circle when the radius is measured up on the circumference. It is approximately 57.3 degrees. Multiplying, the order of the last two terms is still in the billions. If we consider the mass of the Earth also, we are facing an unfathomable amount of angular momentum, at least in our meager human units.

On the next level, Earth has an even bigger angular momentum with respect to its motion around the Sun. The radius of rotation is then the orbit of the Earth around the Sun, about 150 million kilometers, or 150 billion meters. This amounts to $R^2 = 2.25 \cdot 10^{22}$. Even though the angular velocity will be much smaller because now one year is required to make a full circle, hence $\omega = 2\pi / (365 \cdot 24 \cdot 3600) = 0.0000001$ radians per second, the order of the last two terms is still around 10 to the power of 15, or a thousand trillion. Multiplying that again by the mass of the Earth and now we are contemplating a truly out of this world angular momentum.

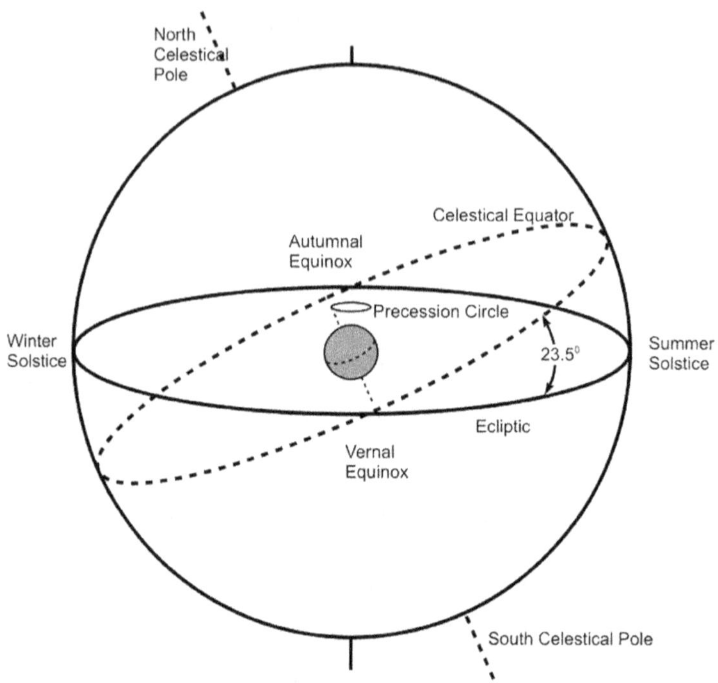

The wobbling Earth

Conservation of this angular momentum in space, lacking any conventional friction, is simple. The angular momentum of the Earth moving around the Sun is balanced by the gravitational pull of the Sun, and to a lesser extent the Moon and other celestial objects. The result is the precession phenomenon, the wobbling of the Earth's axis shown on the above figure. It is due to the fact that the combined gravitational pull on Earth is at an angle with Earth's orbital plane, tilting it at an angle of 23.5 degrees.

The tilted position, contributing to the seasons on Earth, was already noticed in the antiquities. Egyptian and Babylonian observations a couple of thousand years apart reported different locations of the vernal (spring) equinox in the sky. Vernal and autumnal (fall) equinoxes are the points on the celestial equator where the path of the Sun crosses it. The celestial equator is the projection of Earth's equatorial plane to the celestial sphere.

The Vernal equinox occurs on March 21st, that is the day when Sun is directly overhead at the Equator and the day and the night are of equal length. The Babylonian records indicated the location of the vernal equinox to be in the Aries constellation while the Egyptians noted it as being in the Lyra constellation. Hipparchus, in 100 BC noticed this discrepancy.

Hipparchus lived on the island of Rhodes, another Greek scientist of those times whose life is less known but whose contributions survived the millennia. He computed the length of the precession cycle by observing the fixed stars. His computations resulted in a rotation of about one degree per century, amounting to a full cycle of approximately 36,000 years. While his interpretation of the observations was correct, he was unable to explain them without the gravity concept that came almost two thousand years after his time.

It is stated in various sources that the phenomenon was at least observed, albeit not understood, by others a long time before Hipparchus, notably the ancient Indians, the Egyptians and the Mayans. According to certain interpretations of ancient Hindi texts, the full revolution was believed to be about 25,000 years translating their units.

More interestingly, some scholars posit that at the time of the building of the three great pyramids on the Giza Plateau the vernal equinox was pointing to the Orion group. The arrangement of the three pyramids is apparently in agreement with the positioning of the three stars creating the Orion figure's belt. Considering the long, several thousand years of history of the Egyptian culture, it is very probable that their celestial observation records were carried over the generations and the changing equinox may have been noted.

The full precession, we know now more accurately, takes about 25,760 years during which the Earth completes a full wobble. It takes about seventy-two years for the precession angle to change one degree, hence Hipparchus was off by about 30 years. We now fully understand the precession phenomenon and also the reason for Earth's reluctance to give up its axis of rotation. Remember the childhood spinning toys, the buzzers with the pull string or the simple tops that delighted kids for about a hundred years all over the world? It was magical when we tried to tip it over and it stabilized itself after some wobble and kept on spinning. Well, the same magic is at work with the Earth, called the gyroscopic effect. We are all living on a huge celestial gyroscope.

The gyroscope effect has been known for about 200 years and was demonstrated with various mechanical components, spheres, disks and cylinders. Such objects when spun around their axis of symmetry at a high speed, tend to maintain their state of spinning when mildly disturbed. This effect is also due to the angular momentum. The larger the angular momentum (the faster the object is spinning) the larger the inertia of the object is against changing its orientation. Hence a gyroscope is maintaining its orientation despite the external force's attempt to topple it over.

The hero of our chapter, Foucault himself is credited with giving the name to the instrument. Further inspection of the behavior of the gyroscope reveals other intriguing forces resulting from the rotational phenomenon.

10

TORSIONAL EFFECTS

The important discovery of Gaspard Gustave Coriolis in the early 19th century opened up an entirely new understanding of the rotating phenomenon. Coriolis, a mathematician, made interesting observations when working as an engineer. He realized that some of the known mechanical laws do not accurately describe certain phenomena when they are viewed in a rotational frame of reference.

Coriolis was born in 1792 in Paris, the year the monarchy was abolished in France. He graduated from France's premiere school, the Ecole Politechnique, and went to work in the engineering corps for several years. After paying his dues and gaining invaluable insight into some practical problems, he returned to his alma mater to teach applied mechanics.

He published a paper titled *Sur les équations du mouvement relatif des systémes de corps* in the early 1830s. The paper dealt with the equations of relative motion of a system of bodies. It was not about the rotation of Earth or the atmospheric consequences of "his" force; those came later by others. Incidentally Coriolis was investigating the operations of water wheels, the ancient equipment we met earlier, and the transfer of energy between its various stages.

He found the existence of a force acting in a plane perpendicular to the axis of rotation and proportional to the product of the mass (m), velocity (v) of the moving body and the angular velocity ω as $2\,m \cdot \omega \cdot v$. He found that if the plane of the motion is not perpendicular to the axis of rotation but at an angle α then the magnitude of the force becomes $2\,m \cdot \omega \cdot v \cdot sin(\alpha)$. This force is now the Coriolis effect.

In the atmosphere, air flows from high pressure areas toward the lower pressure areas, just like heat moves to lower temperatures. When the air flows in the horizontal plane at a certain latitude angle α on the rotating Earth, it is also influenced by various levels of the force, now named after Coriolis.

Since hurricanes are also the result of air moving from higher pressure to lower, they are also influenced by the Coriolis force. The hurricanes of the Caribbean dominantly rotate counter-clockwise and the cyclones of the southern hemisphere clockwise. This is due to the that the vector of the axis of rotation points to North on the Northern hemisphere, but to the South on the Southern Hemisphere.

There appear to be neither cyclones, nor hurricanes in regions close to the equator, as from the above equation it is visible that for very low altitude angles (α) the Coriolis force is small due to $sin(\alpha)$ being very small. On the other hand, the arctic regions, where the force is large as $sin(\alpha) = sin(90) = 1$, the rotational turbulence is extraordinary.

Let us now jump to the late 19th century when a German academic research group conducted gravity measurements aboard a ship on the North Sea. The resulting data was made available to researchers in Europe and a most intriguing observation was made by scientists. They noted that the measurements were repeatedly different whether they were made by moving east or west over the same point. This phenomenon was very noticeable and confounded the scientists at the time.

Lóránd Eötvös, scion of a Hungarian noble family, was born in 1848. It was the famous year of his nation's history when Hungary started another ultimately loosing fight for secession from the Austro-Hungarian empire. As usual at the time in Hungarian noble families, Eötvös entered law school aiming at becoming a politician. His heart was, however, set on natural sciences and after some period he switched to study such at the University of Heidelberg in Germany. He was educated by such luminaries of science as Gustav Kirchoff of the circuit law fame and Hermann Helmholtz, the discoverer of the conservation of energy principle.

After graduation he returned to Pest and obtained a lecturer position at the University of Sciences, now bearing his name (and the alma mater of your humble author as well). His field of expertise was experimental physics and he had a strong interest in gravitation.

Eötvös recognized that since the plane of the Coriolis force is perpendicular to Earth's axis, it also has a component perpendicular to the plane of the motion at various latitudes. This force is the Eötvös effect in the form of $2\,m\cdot\omega\cdot v\cdot\cos(\alpha)$.

He concluded that the phenomenon encountered by the German scientists was due to this perpendicular component of the Coriolis force arising at the particular latitude. The effect is since known as the Eötvös effect and the horizontal plane component of the Coriolis force hence also now called the Coriolis effect, as shown on the following figure.

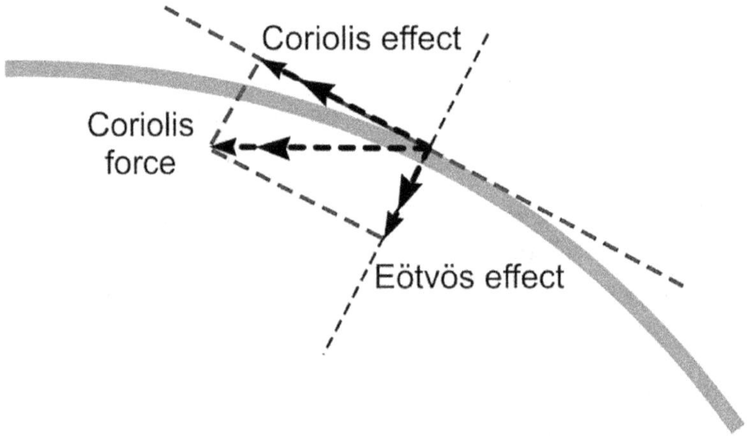

Coriolis force components on Earth

The value of $\cos(\alpha)$ is one when the angle is zero. That corresponds to the equator, hence the effect of the vertical component of the Coriolis force is at maximum there. On the converse, there is no Eötvös effect at the pole on account of the value of $\cos(\alpha)$ being zero when the angle is 90 degrees. Combining this with the directional dependence of the Coriolis force one can conclude that moving east the direction of the effect is away from the surface of the Earth. Moving west, the Eötvös effect points into the surface of the Earth, adding to the force of gravity.

The above observations are appropriate for the northern hemisphere, where a counter-clockwise rotation fits the orientation of Earth's rotation vector. This is established by observing the rotation while looking into the axis. On the northern hemisphere, looking up toward the pole, the rotation of Earth is counter-clockwise. The force would be the opposite in the southern hemisphere, since looking toward the south pole Earth's rotation is clockwise.

Therefore the weight on an eastward moving ship on the northern hemisphere is lessened by the vertical component of the Coriolis force. We are weighing less when we move eastward. Now with some humility we'll have to admit the fact that in the scale of our personal weight this is not measurable, so there will not be diet plans proposing weight-loss by constantly traveling eastward. Nevertheless, it is another interesting effect of the rotational phenomenon in our lives.

The actual measurement of the effect is done by a heavy object suspended on a string. The required suspending force is exactly the same as the gravitational force, i.e the weight of the object. By the well known Hooke's law the force in the spring is proportional to its lengthening. It is rather simple to measure the length of a spring to a high accuracy, hence the measurement differences between the east and west bound paths were noticeable.

It is perhaps time to put the Eötvös effect in a quantitative perspective. At the latitude angle of 60 degrees, the cosine becomes one half and the vertical Eötvös effect acting on an object of mass m will simply be $2\,m \cdot \omega \cdot v \cdot 0.5 = m \cdot 0.000073 \cdot v$, where we used the angular velocity of the Earth established earlier in radians per second.

This force, along with the force in the spring counters the weight of the object. Assuming that a 100 kg object is moving eastward with 10 meters per second, the upward Eötvös force becomes approximately 0.073 Newton. This will lessen the force in the spring, hence shortening its length. On the contrary, moving to the west, the vertical force is added to the weight and therefore further lengthening the spring.

The difference amounts to about a third of an ounce in weight, hence the measured length of the spring between the two paths is detectable by a delicate instrument. The effect diminishes as we move toward the poles and increases toward the Equator.

Eötvös' explanation was not immediately accepted. The original German research team, however, repeated the experiment in 1908 on the Black Sea, somewhat closer to the equator. Simultaneous observations were made in two boats independently. The results were the same after adjusting to the latitude difference.

There are some other interesting effects of the Coriolis force. Among them is the different whirling of the water in a sink between the hemispheres. Specifically, on the northern hemisphere, just like the hurricanes, the water vortex is counter-clockwise while on the southern is the contrary. Of course in a real sink the symmetry and scale are influencing the phenomenon.

Australian researchers, for whom this was a topic of valid scientific interest, used a symmetric cylindrical tank, filled with water to a significant height and let it settle until it was stationary. Then slowly opening a whole in the center of the bottom of the tank, they found that after the initial turbulence caused by the opening, the steady state water vortex became clockwise, as predicted by Coriolis force.

11

PERIODIC TIDES

Imagine yourself sitting on the beach and watching the approach of the waterline on the shore. There is some calming and at the same time disturbing aspect of it, the wave noise could lull your senses but the relentless coming of the water makes you wonder about the forces that generated it.

The periodic nature of tides was subject of intense speculation throughout humankind's history, especially since the early concentrations of advanced cultures were mainly around the Mediterranean. Hence, there is a wide-spread history and many notable scientists involved in the explanation of the tidal phenomenon.

A lunar relation of tides was recognized very early, a description by the Babylonian Seleucus already attributed the tides to Moon's cycles. There are more records indicating such recognition in other sea-side cultures, for example by the Greeks and ancient Chinese several centuries BC.

As revolutionary as Galileo's thinking was in his time, he was not infallible. He was asked by a high ranking member of the Catholic church to explain the tides. He attributed them to the rotation of Earth around its axis and the resulting sloshing of the ocean's waters. This of course contradicted the observation of two daily tides on the Adriatic coast of Italy, specifically in Venice where the topic was of immense interest. He dismissed this as a result of the local behavior of the Adriatic, due to its special shape and semi-enclosed nature. He was wrong and the phenomenon is of course due to the gravity pull of Moon as we will see later.

The first organized observations in the form of tidal tables appeared about in the 11th century in China and in the 13th century in England. Both were related to river locations in close proximity to their outlet to the sea; in China the river Qiantang and in England the Thames.

The fact that they all realized the relationship with the rotating Moon did not mean that they recognized the gravitational role of the Moon in it. Newton was the first who tied gravity and tides together, and the French scientist Pierre Simon Laplace provided a mathematical representation for the phenomenon.

Laplace was born about a hundred years after Newton in a small Normandy town into a non-noble family. While serving in the army in 1784, he briefly tutored a young 16 year old cadet from Corsica by the name of Napoleon Bonaparte. Despite Laplace having an apolitical personality at the time, his life was endangered during the tumultuous French revolution at the end of the 18th century. He and his family narrowly escaped the revolting Paris mobs and sought refuge in the countryside.

The tidal effect is the result of a rotational and gravitational tug of war. Earth's acceleration of gravity on the surface of Earth is slightly modified by the rotating Moon's own gravity. The difference between Moon's effect on Earth and Moon's effect on a body (for example the water in the oceans) on the surface of Earth is the tidal acceleration. The lunar tidal acceleration is about 10^{-7}, or about tenth of a millionth of Earth's own acceleration of gravity. While that seems almost negligible, it is not in the grand scheme of things as attested by the tides we see daily.

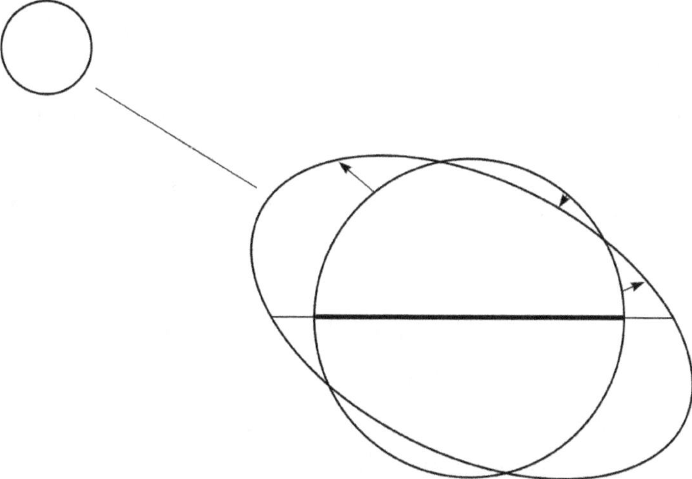

Tidal waters on Earth

In analyzing the tides, Laplace considered Earth to be fully covered with water. This is not as far fetched as it sounds, after all, almost two thirds of Earth's surface is covered with oceans. In the above figure the small circle on the left represents Moon, while the large circle represents Earth's surface. The ellipse represents the deformed shape of the water surface of Earth, due to the gravitational pull of Moon. The arrows represent the direction of the moving water surface. They show that in some areas the water level decreases, while in others increases.

The increasing water level on the side that is closer to Moon makes sense, after all, that is where Moon's gravitational pull occurs. This side is called the sub-lunar side for obvious reasons. The other side, called antipodal, begs for an explanation. Why would be a bulge on that side also? The explanation for that is because Earth is also being pulled by Moon toward it and to a lesser extent that the water on that side, therefore there will be a higher water level there as well.

This is also the reason why we get two tides a day. One occurs as our sub-lunar tide when Moon is up and one we get as an antipodal when it is down. Tides therefore occur with a reasonable repeatability, we get high tide twice a day about 12 hours and 25 minutes apart. There is half that time between the high tide and the low ebb.

There is a strong variation of the tides depending on the relative position of Moon with respect to Earth. Specifically, the distance between them changes by as much as 50,000 kilometers during the duration of a month. Once a month, when Moon is the closest to Earth in the so-called perigee position, the tidal forces are the highest. On the other hand when Moon is at the apogee, farthest from Earth, the tidal forces are the lowest.

Another variation of the lunar tidal effect is due to the fact that Moon's orbital plane, in which it is rotating around us, is inclined about 5 degrees to the equatorial plane of Earth. Hence Moon, when making one monthly revolution, is making a round trip between a maximum position above the Equator to the same distance in the South. As a result, Moon crosses the equatorial plane twice a month. This effect of Moon's declination is named the diurnal tide pattern.

The tidal effect of Moon also depends on its relation with respect to the Sun. The solar tidal acceleration is about half of that of Moon. This seems contra-intuitive since the Sun is much larger than the Moon but we can prove it by simple arithmetic. Sun is almost 27 million times bigger than Moon, but it is also 390 times farther away. Since the tidal acceleration diminishes by the cube of the distance, dividing 27 million by the cube of 390 yields about one half of the tidal acceleration of Moon. Sun's effect is far from being negligible and sometimes exaggerated by the relative positions of the Sun and Moon.

When Moon is either full or new, it is on the line connecting the centers of Earth and Sun. In the case of new Moon, it is on the side of the Sun and the tidal forces of both Sun and Moon are combined. These are the highest tides, sometimes called the spring tides, although they do occur in the other seasons as well. When Moon is on the opposite side of the Sun, at full Moon, the tidal forces are also combined and spring tides occur. The effect is now produced by the antipodal tide scenario, but both high tides are about the same.

When Moon is in the intermediate phases, the so called neap tides, or the low high tides occur. Their cause is that the tidal forces of the Sun and Moon are perpendicular to each other. Wherever Moon is raising the water level, the Sun is lowering and vice versa. It is not an exact cancellation, since the tidal acceleration of Sun alone could produce only about one half of Moon's tidal variations. Hence these are low high tides, but still higher than the average water level.

Laplace created a mathematical model of the tidal phenomenon as function of the average assumed water depth, and the latitude and longitude of the location. He also considered the special effects caused by Earth's rotation, the centrifugal and gravity forces, and derived a set of three partial differential equations. The three solutions of the equations were the temporal rate of change of the tidal elevation (tidal surface height), and the acceleration of tidal water in latitude and longitude directions at a certain global location.

12

MOON LANDINGS

It was science fiction at the time Jules Verne wrote his famous "From the Earth to the Moon" book in 1865 that of course became reality about a hundred years later. In fact the ways of accomplishing it, by building a humongous gun and shooting a cabin containing the travelers to the Moon, may have been inspired by Galileo's work on the projectiles, and ultimately was very close to the actual realization of the trip about a hundred years later.

Let us consider the issues of such travel now from gravity's point of view. The fact that there were certain positions in the Sun-Moon system where tidal forces canceled out brings us to an extremely interesting and practically useful gravitational scenario first recognized by Lagrange.

Giuseppe Luigi Lagrangia was born in Italy in 1736 to Italian parents, but due to French heritage on his Father's side, he changed his name to Joseph Louis Lagrange. He spent half of his adult life in Berlin where he produced significant results in the area of mathematics, classical mechanics, but also luckily for our topic in celestial mechanics. Some of these results were acknowledged by the French Academy of Sciences and he moved to France in the last decade of the 1700's just in time for the French revolution.

During those years his path also crossed Napoleon's who was appreciative of Lagrange's science. He gave him the Legion of Honor and even made him a Count. Lagrange became the first professor of mathematics at the newly opened École Polytechnique, France's premier school of higher learning.

Lagrange's analyzed the gravitational relationship between Earth and Moon. This was described by a fifth order algebraic equation whose solutions were five locations where the gravitational forces of the two bodies are in a special relationship and fully or partially cancel each other out. These points are now called Lagrange (sometimes libration) points and their location is shown in the figure below.

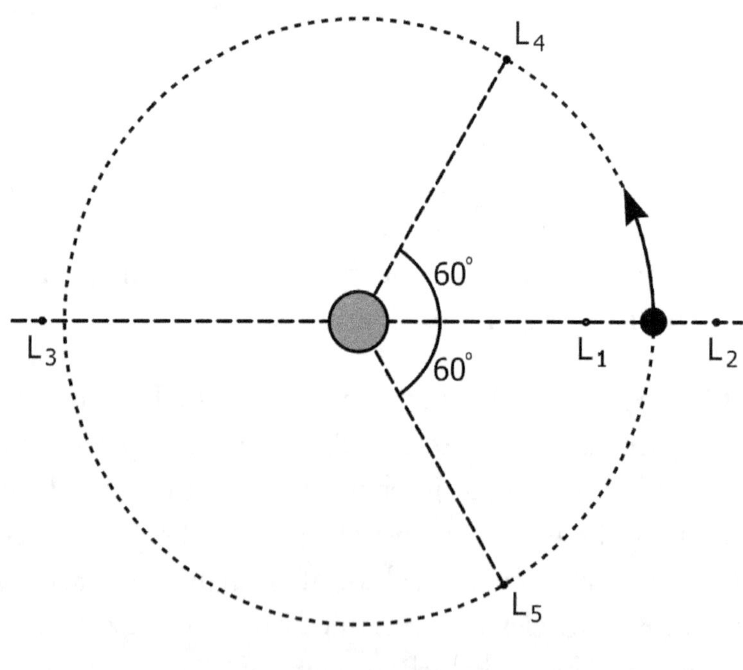

Lagrange points

One such point is called L_1 that is the point on the line connecting the centers of Earth and Moon at the location where the gravitational forces are opposite and actually cancel each other out. The importance of that is in connection with the rotational periods. As we saw it in an earlier chapter the circumferential velocity and the gravitational force are in a delicate balance. An object located in the L_1 Lagrange point has the same rotational period as Moon.

More importantly, an object traveling from Earth to the Moon would change gravitational hosts at that point. If a spacecraft reached this position, from here Moon's gravitational pull would be its governing force. In other words, the object would simply fall onto the Moon from just beyond this point.

Gravity is the main reason for the difficulty of making aircraft leave Earth and reach space. For example, it requires tremendous energy to lift a space shuttle off the ground. The velocity required for an object to escape the confines of the gravity field of a planet is, however, independent of the mass of the object. The bigger the mass the more energy is needed to accelerate the object to the escape velocity, but the required velocity is the same.

When an object is already in an orbit around the Earth, it has a smaller escape velocity. The next figure demonstrates potential orbits around Earth. The ellipse denoted by E and the circle C are orbits of an object with less than the escape velocity. The object will remain in the gravitational hold of the Earth. One could consider the circular orbit being that of our Moon. When the object reaches or exceeds the escape velocity, it will leave the Earth on either a parabolic (P) or a hyperbolic (H) trajectory, respectively.

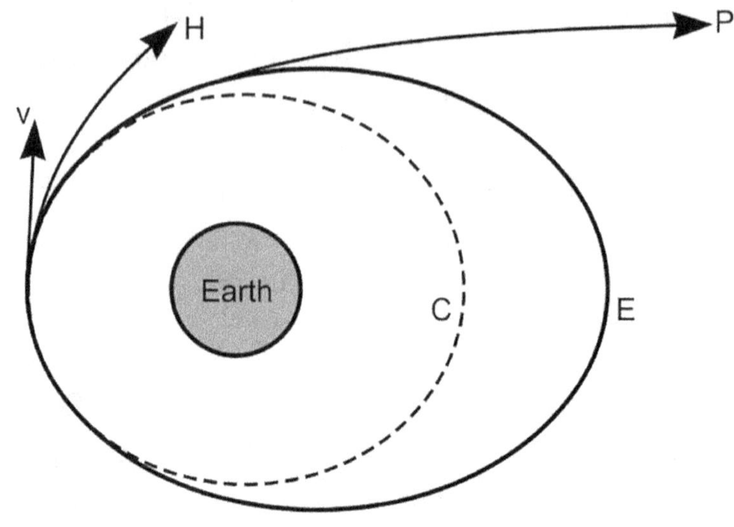

Escape trajectories

More specifically, the escape velocity is the initial speed required from a stationary position and as such it is measured at the surface of Earth. On average, the escape velocity from Earth is about 11.2 kilometers per second. That velocity is about ten times the speed of a bullet shot from a good rifle. Clearly, we need an extremely good rifle to shoot something out of Earth's gravitational field.

The above is actually the second escape velocity. The first one is the velocity of 7.9 km/sec to put a satellite into Earth's orbit. There is also a third escape velocity of 42.1 km/sec that would enable a craft to leave our solar system. Finally, the fourth escape velocity is at 320 km/sec that we need to surpass to leave our galaxy, the Milky Way.

The value of escape velocity assumes that the object is shot up in a vertical direction from Earth and needs to fight head on against the gravitational field. Earth itself, however, could help us in this endeavor. Since Earth's angular velocity due to spinning around its axis at the Equator is about 0.465 kilometers per second, if we launch our object to the east horizontally, we only need about 10.735 kilometers per second escape velocity. It does not seem much of a difference, but when considering the energy needs of practical objects (certainly larger than a bullet), this is significant. This is the reason the launch site of the US space program is as south on the continent as possible: Cape Canaveral.

Cape Canaveral was of course where the historical Moon landing expeditions were launched from. The first such operations were so-called hard, and later soft landings. This distinction is due to the fact that after reaching the L_1 point the challenge becomes to reverse the gravity force of Moon for landing.

Hard landings mean essentially falling to the surface of Moon and that is somewhere in the neighborhood of 2.65 kilometers per second, some 9,500 kilometers per hour. Building a structure retaining its integrity after such a crash is very difficult. Some of the hard landings were successful in taking some photos, but the craft were destroyed.

In the second half of the 1960s several American spacecraft were successfully soft landed. The Surveyor probes were able to gather soil composition data and transfer back, electronically of course. Then the issue of actually bringing something back physically arose. Another fight with Moon's gravity came up now in the opposite direction.

Moon's escape velocity due to its smaller mass is significantly smaller than that of Earth's, but still a formidable 2.38 kilometers per second. This seems like a problem already solved, after all, if we were able to escape Earth with the space craft, escaping Moon should be easy, shouldn't it? Well, the difference is that the craft landing and ultimately taking off from Moon is not the same as the one that left Earth. As we saw in the old Apollo films and the new films of space shuttle take-offs, leaving Earth requires a multi-stage rocket whose pieces are discarded in the process.

The Moon landing craft included a dual use rocket that on arrival decelerated and on departure accelerated the craft to overcome the escape velocity. Let us now assume that we are off the surface of Moon and on a return trajectory. The fight with gravity is not over because we need to overcome again the gravity of Earth in an opposite sense and prevent the destruction of the returning capsule. That is made somewhat easier by Earth's atmosphere since it's resistance itself decelerates the craft, however, it also generates a lot of heat. Hence the landing module, the capsule containing the astronauts, contained heat shields and parachutes to complete the trip.

Moon landings with humans on board were successfully done 6 times during the Apollo program of the 1960s. Apollo 11 was the first, with Neil Armstrong being the first human to set foot on Moon on July 20, 1969. A Saturn V rocket launched from Cape Canaveral on July 16th carried the command module Columbia and lunar landing craft Eagle into orbit. Columbia was actually named after Jules Verne's giant cannon.

Neil Armstrong's first steps on an extraterrestrial surface were watched by more than half a billion people worldwide. The extraordinary feat was repeated by Apollo 12, 14, 15, 16 and 17 with the last human on Moon being Gene Cernan on December 11th, 1972. Apollo 13 was of course the so-called "successful failure" mission commemorated in the movie of the same title. The crew was forced to take refuge in the landing module due to an onboard explosion and returned safely to Earth after taking a loop around Moon.

Returning to the Lagrange points, the second point, L_2 lies on the same line but just a bit outside of the orbit of Moon. The Earth-Moon L_2 point could be a good location for a space based observatory to see Moon's far side. The third point, L_3 is on the opposite side of Earth. Finally L_4 and L_5 are located at 60 degrees from the line connecting the two bodies and again slightly outside of the orbit of Moon.

Naturally, Lagrange points exist in connection with any pair of celestial objects. The practical importance of such locations is enormous. For example, the Sun-Earth L_1 point is a prime location for space based observatories of the Sun. In fact that is where the SOHO satellite (Solar and Heliospheric Observatory) is located.

The L_2 of the Sun-Earth system is also a good location for a space based observatory. It is about 1.5 million kilometers outside of Earths orbit around the Sun. It also has the desirable characteristic of having the same rotational period as Earth, but sort of shielded by Earth from the Sun. It was the location of the Kepler space telescope launched in March 2009. The telescope's first success was in 2011 by the finding a star with a habitable planet, now named Kepler 22B about 600 light years away. Kepler was retired after ten years and the James Webb space telescope is now in orbit around the Sun-Earth L_2 point.

The Sun-Earth system's L_3 is on the other side of the Sun and somewhat less interesting for humankind. That is also less desirable for practical use because the gravitational effects of the other planets will come into consideration and make it instable. It was part of the science fiction lore for some years as the possible location of the twin of Earth's, an appealing concept to lovers of symmetry theories.

The L_4 and L_5 points' balance is even more interesting. As visible on the figure, they are forming an equilateral triangle with the two celestial bodies of the system. The balance there is such that the ratio of the gravitational forces from the two objects is the same as the ratio of their masses. The objects placed in these points will be in synchrony with the system. The L_4 point is orbiting 60 degrees ahead, the L_5 point is 60 degrees behind the smaller body, almost like a celestial front and a rear guard.

These points are very stable and can contain objects with relatively large mass. Nature itself recognized this fact. The Sun-Jupiter system's L_4 and L_5 points are occupied by two asteroids, called the Greek and Trojan asteroids, respectively. Hence sometimes those points are also called the Trojan points.

Another Moon related excursion, albeit not resulting in landing, was the September 2011 launch of a pair of spacecraft from Cape Canaveral atop Delta II rockets. The mission was historic in the sense as these were the last of such rockets to be launched.

The mission was to execute gravitational measurements in hopes of finding out the interior composition of Moon. As one of the NASA program managers said "this is a journey to the center of the Moon". The mission was looking to find out whether Moon has an inner core. Does it have fluid around it and an Earth-like mantle, or is it just a big piece of round rock?

The mission was called GRAIL (Gravity Recovery And Interior Laboratory) and the craft were aptly named by a group of 5th graders in the elementary school of Bozeman, Montana to Ebb and Flow. The craft used a special trajectory that required lower energy but it took longer, almost about 3 months to reach Moon. In contrast, the Apollo missions reached Moon in only about 3 days.

The craft reached Moon's orbit on New Year's day in 2012. After some adjustments they started rotating around Moon with a radius of approximately 30 miles and an orbital period of about 2 hours. They rotated in tandem while maintaining their distance to an accuracy of less than ten thousandth of an inch.

Since they were in a distance of 120 miles from each other, they encountered gravitational fluctuations at different times and that enabled accurate relative measurements. These variations provided information about the interior constitution of Moon. After succesful measurements, their energy exhausted, there were crashed into Moon in December, 2012.

13

WHIRLING HOLES

Black holes are singularities humorously described as the result of God trying to divide the universe by zero. The German scientist Karl Schwarzschild proved the possibility of their existence as a consequence of Einstein's equation in 1916.

Schwarzschild was a child prodigy already publishing papers about celestial mechanics at the age of 16. He was born in Frankfurt, Germany in 1873 and conducted studies in Strasbourg and Munich, earning a doctorate at the age of 23. After several years of heading an observatory in Vienna, he was appointed to be the director of the observatory at the famous university of Göttingen, home of such luminaries of science as Gauss.

After a decade and a half of fruitful scientific research, the outbreak of World War I changed his life. He joined the army and continued doing research work while on the Russian front. Incidentally that included finding an exact solution to Einstein's gravitational equation. Upon his return he sent his solution to Einstein and soon after he died as the result of a disease he developed on the front. He was honored by an asteroid named after him: 837 Schwarzschild.

Einstein's gravitational equation is still extremely difficult to solve exactly, in fact Einstein never solved it exactly. The 10 solutions of the equation specify the ten distinct components of the geometry tensor of the gravity field and describe the geometry of the curved space.

Einstein's geometry tensor was a symmetric matrix of 4 rows and columns. The four dimensions were the 3 spatial directions and time. The symmetry results in 10 unique terms, four on the diagonal and 6 above, these are the 10 solution parameters of Einstein's equation. Hence Einstein's tensor equation may also be described in 10 nonlinear, partial differential equations, the same types Laplace used and that are somewhat beyond our mathematical level here.

Schwarzschild's exact solution was assuming a spherically symmetric but not rotating mass generating the gravity field. His solution produced a special term which is now called the Schwarzschild radius. The formula of the term depends on Newton's gravity constant, the governing mass and the speed of light.

The meaning of this radius is that if an object's geometrical radius is smaller than its Schwarzschild radius, it is a very high concentration of gravity, the aforementioned black hole. The surface of the sphere with this radius is the so-called event horizon of the object. That is the neighborhood of the black hole from beyond which there is no return. Neither for mass, nor for light, hence the name.

Einstein himself was not too pleased about this possibility, after all, his approximate solution did not yield this scenario.

Nevertheless, the presence of such objects is now widely accepted as being massive stars that collapsed under their own gravity. It is believed that the center of our own Milky Way galaxy is a giant black hole.

The black hole at the center of our galaxy is sometimes called a galactic glutton as it seems to have an insatiable appetite for material. The sizes of black holes are usually classified by their density. That is computed as the amount of mass in the black hole divided by its Schwarzschild radius. The amount of mass in the black hole of Milky Way is estimated to be 2 million times the mass of the Sun. Its density is assumed to be larger than the $1.8 \cdot 10^8$ mass unit per volume boundary, classifying it as a super massive black hole.

The event horizon for the black hole at the center of our galaxy is about 36 light hours, or close to 40 billion kilometers, a very formidable distance even in galactic terms. Popular science films depict the fate of an object going through the event horizon as being stretched out like a long spaghetti.

There are many black holes around us. One super massive black hole named NGC 3842 (New General Catalogue of nebulae and clusters) is located about 320 million light-years from Earth in a cluster of galaxies in the Leo constellation. It is about 9.87 billion Sun masses. Then there is another one in the Coma constellation. This is about 336 million light-years away with almost 20 billion Sun masses, and it is the present record holder.

Black holes are not only a dead end of collapsed celestial objects, in fact they are supposed to give us a glimpse of the formation of galaxies as well. Some scientists suspect that the center of every galaxy has a black hole and they are somehow related. The research is now executed with orbital observatories, such as the Hubble and Webb space telescopes.

To put the concept of the Schwarzschild radius into human perspective, for the Sun it is about 3 kilometers and we know that the Sun is much bigger than that. Similarly, the Schwarzschild radius for Earth is only about 9 millimeters and we know that its geometrical radius is about 6,400 kilometers. We are safe from Earth or Sun collapsing into a black hole soon.

The concept of the event horizon and Schwarzschild's radius brings another apparent contradiction with prevailing scientific beliefs. The big bang hypothesis proposes that there was an extremely small, but very dense kernel of primordial material that exploded and resulted in the material in the universe. The theory proposes that the bang happened in a time frame of only about 10^{-36} seconds and during that time an expansion of volume by a ratio of 10^{78} occurred.

These numbers imply that the universe at the beginning was much smaller than its event horizon and therein lies the contradiction. Einstein's theory of gravitation via its mathematical consequence of the black hole in Schwarzschild's solution proposes that nothing can escape from the region inside of the event horizon and that seems to fully contradict the big bang theory. Since the current size of the universe is of course much larger than its event horizon, there must have been a process to get material out of the original event horizon.

One way out of this contradiction is by saying that Einstein's gravity theory simply does not apply in these circumstances.

The primordial environment was an extremely hot (perhaps millions of degrees of Celsius) cloud of elementary components of material, whose nuclear interactions of extremely high energy overrode any other physical considerations.

Another idea is that the big-bang immediately resulted in rotation as well. After all, it is an everyday occurrence of celestial objects and rotational black holes could be the result of spinning stars collapsing under their own gravitational field. However, in Schwarzschild's solution the spherical object was not rotating.

An exact solution to Einstein's equation for such scenario was produced in 1963 by Roy Kerr, a New Zealand mathematician. He extended Schwarzschild's model by allowing the rotation of the mass. This solution is considered to be the most important solution of the equation since it is the best fitting to our universal observations.

Roy Kerr, born in New Zealand in 1934, was also a mathematical prodigy like Schwarzschild. He completed his studies at the University of New Zealand as a teenager and had to wait for his formal graduation papers until he reached 20 because of regulations. He earned his doctorate at Cambridge University in 1959 with a dissertation already focused on Einstein's equation and before he turned thirty he presented his solution.

Kerr also received his share of medals for his outstanding work in theoretical physics, and ultimately the Companionship of the New Zealand Order of Merit in 2006 "for his services to astrophysics".

Kerr's solution also produced a black hole, but with two singularities. The inner singularity was a spherical event horizon, similar to Schwarzschild's. The outer singularity was a surface of a flattened sphere, not unlike our Earth. Between the two singular surfaces the material would be whirling with a gradually diminishing rotational speed. The rate of change of this rotation speed depends on the angular momentum and mass of the originating star. It is now believed that the black hole at the center of our galaxy is also whirling.

Whirling black holes may be created when a rotating object's angular momentum reaches a critical minimum that is a function of its mass, rotational radius and angular velocity. This is along the line with the reasoning we followed earlier in connection with Moon's motion. The angular momentum held Moon speeding along on its orbit balanced by the gravitational force. As we mentioned there, if Moon would loose its angular momentum, it would fall to Earth.

Our Sun, formed about 4.57 billion years ago, is an interesting example since it also has a variable rotation. At its poles it rotates in 35 days and at its equator in 25 days. Its estimated mass is about $2 \cdot 10^{30}$ kilograms and accounts for almost 99.9 % of the mass in the solar system. The diameter of the Sun is 1,392,000 kilometers and its angular momentum is approximately 5 times the critical minimum.

It appears that we are on safe grounds with Sun and it will not become a whirling black hole in the near future, although the current science is that black holes are inevitable results of collapsing stars. There is a rather far fetched explanation of the possibility of white holes. A white hole, as the name indicates, would be the opposite of a black hole. Material could not enter the white hole, but matter including light could leave it. The concept hinges heavily on the time component of the four dimensional space used in Einstein's theory and sort of proposes a temporal progression.

In this theory, the white holes are the past and black holes are the future. The material of a past white hole could explain the big bang hypothesis, and since that happened many billions of years ago, it fits the past history scenario.

The white hole-black hole scenario is even more attractive in connection with the theories proposing the cyclically expanding and contracting nature of the universe: the big bang being followed by a big crunch. These are speculations, of course, hence we will continue with observable celestial cycles.

14

COSMIC CALENDARS

The Mayans opened up their eyes to an even wider cosmic frame and created a really long calendar that reaches into our time. It appears that their long calendar was started in 3114 BC. Obviously this date was computed a posteriori since our current time keeping method had not been defined until the middle ages as we'll learn in the following chapter. This is the calendar whose apparent end in 2012 was considered to be the day of Apocalypse by some. As we know, those doomsday predictions were unfounded and fueled by the misinterpretation of the long calendar.

Celestial cycles have been observed by the Mayans for millennia and there is no reason to assume that this is their first long calendar. As the new cycles have replaced the old ones, the calendar may also have been re-cycled by the ancient priests. After all the recording material was scarce and very expensive - another component of modern life that may have had thousands of years of history.

There is no question about the fact that some interesting celestial events took place in 2012. There was a solar eclipse and a transit of Venus. There was a less definitive event of the Sun eclipsing the center of the Milky Way.

The latter is a bit murky because the center of the Milky Way is somewhat of a relative concept, based on our perception that is skewed by the fact that we are in it and it is a rotation system. The center is a whirling black hole and we have a hard time comprehending its existence, let alone defining the meaning of that celestial scenario. But, three is a special and magical number, and predicting three major celestial events in a year certainly makes it special, indeed. So the calendar ended, that is a fact. Let us see the calendar in more detail to understand the possible reasons of its ending.

The Mayan calendar was of course written with their hieroglyphs that were only deciphered in the 19th century. Its cycles were based on a specific, largely base 20 place system. Their digits, separated by a symbol and represented here by a period, designated the multiplier of the appropriate place value. Just like in our decimal (i.e. base 10) system 15 stands for $1 \cdot 10 + 5 \cdot 1$, the value of 2.5 meant 45 days ($2 \cdot 20 + 5 \cdot 1$) in the Mayan's base 20 system. However, in their desire to reconcile their counting with the years (albeit approximately) the base was switched between the second and third positions to 18 and resumed thereafter.

Hence the place values in their system were: 1; 20; 360; 7,200 and 144,000. These place values had designated names: Kin, Winal, Tun, Katun, Baktun. A date of 1.0.0.0 in their long calendar was 7,200 days, or one Katun. Similarly, the date of 1.0.0.0.0, or one Baktun amounted to 144,000 days, or approximately 394.3 solar years.

To put these numbers in perspective, let us tabulate some years of the current (or as it is called last) cycle of the Mayan long calendar.

We will convert some specific dates. They may be off by a few days for reasons of various adjustments of the calendars during the millennia, as will be seen in the following chapters. The number of elapsed days from the calendar start to certain dates will show that the end date can easily be explained by their number system.

Modern date	Mayan date	Elapsed days
08/11/3114 BC	0.0.0.0.0	0
12/06/2720 BC	1.0.0.0.0	144,000
12/07/1143 BC	5.0.0.0.0	720,000
01/01/01 AD	7.17.18.13.1	1,137,141
03/09/830 AD	10.0.0.0.0	1,440,000
09/19/1618AD	12.0.0.0.0	1,728,000
12/12/2012AD	13.0.0.0.0	1,872,000
03/09/2410AD	14.0.0.0.0	2,016,000

As the table above shows, in late 2012 we arrived at the conclusion of a great cycle of one Baktun, denoted by 13.0.0.0.0 in their long calendar. There might be some minor discrepancy in the actual date since there is no way to know the exact date of the cycle's starting month, after all those concepts did not exist before. The actual number of days of 1,872,000, puts the start somewhere in August of 3114.

It is quite rational to assume that the Mayans simply started another calendar cycle on the largest place value (Baktun) at that time and that is the reason why it ends in 2012. The fact that it is the end of the 13th great cycle, however, may have some superstitious, more modern relevance, although the number achieved its notorious nature only much later in the middle ages.

As an additional assurance of our future, note that there are seven more Baktun cycles before we would reach 19.19.19.17.19, the apparent end of their known number system. That would be more than 8 thousand solar years, well in our future. Furthermore, their system did not even end there.

They had four more, lesser known, higher order (by ratios of 20) place values, named: Piktun, Kalabtun, Kinchiltun and Alautum, enabling one to write numbers as large as 19.19.19.19.19.19.19.19.17.19 This is almost as large as 20 to the 10th power (ignoring the minor glitch of switching to 18 once) and this number is astronomically large, measurable only on a cosmic time scale as it far exceeds even the estimated age of our universe.

Some interpretation of the start of the last calendar states that the year 3114 BC was the end of the previous world, probably civilization, in Mayan beliefs. It may also have been the date of Kukulcan's founding of their new civilization. Whether one Baktun cycle of 5125 years is fitting to some of their historical recollections is not proven, but it is certainly noticeable that their civilization did not last a full great cycle and became extinct, quite earlier.

There are some records indicating that the Mayans believed in very large cosmic cycles, called the ages of the Sun. The school of thought called Mayan cosmogenesis credits the Mayans with a deep understanding of our cosmic origins.

The cycles may not indicate the beginning and end of creation, instead they provide the means to measure the apparent cycles of cultures or civilizations. In the Mayan history there were already four worlds preceding the present one that were destroyed by some events. As we recall, the number 5 held significance to the Mayans from their Venus observations.

The end of the 5th Sun was described by the Mayans as the time when the soul of humans will be released from the material confines (i.e. body) to reach the cosmic heights. This is very similar to other cultures' description of ascending to the heavens and may be the main contributor to the apocalyptic visions.

One thing seems very peculiar: 5 cycles of the Maya long count, $5 \cdot 5125 = 25,625$ is a remarkably good approximation of the length of the precession cycle that takes 25,760 years during which the Earth completes a full rotation with respect to the universe.

This coincidence is too close to ignore. It is conceivable that the Mayans somehow were aware of the fact that our present North Star, Polaris was once our North Star before. We now know that the Vega was our North Star around 12,000 BC and it will be again in another 14,000 years or so. From 1,500 BC to 500 AD, the Beta star of the Little Dipper constellation "served" as the North Star. Since the Little Dipper is also visible to the naked eye in the night sky and its North Star position overlapped with the Mayan times, they must have noticed the moving of the position of the North against the stars.

The Mayans kept records of their celestial observations through generations and it is not a big leap of faith to assume that they recognized the rotational precession. Such a big time cycle has certain resonance with the theories about the cyclic expansion and collapse of the universe, and renewal by the big explosion again.

The early recognition of our world being a part of great cosmic and celestial rotational cycles, did not alleviate the need for humans to describe the smaller cycles of our lives. This topic was not without its challenges as we will see it next.

15

SOLAR YEARS

Observing the full cycle of the Sun led the Mayans to embark upon a solar calendar making activity. Their year was comprised of eighteen months with twenty day duration. This still did not correspond exactly to their yearly Sun cycle observations, therefore 5 days long adjustment periods were added at the end of every year. They started their calendar year with the winter solstice, when the Sun is at its southernmost position.

They also created a mid-range calendar of 52 solar years, the so-called calendar round, that may be viewed as an extension of the celestial cycles into a human life cycle. This calendar was possibly maintained by a high priest throughout his lifetime.

The Mayans were not the only civilization observing and relating to the celestial rotational cycles. The Chinese also had solar calendars, dating back as far as 5 millennia. It was a luni-solar calendar, predominantly based on the Moon's cycles. In it, each month started at one hour before midnight of the day of the dark Moon. A year had twelve lunar months which required inserting an additional month every third or fourth year in order to coalesce again with the solar cycle.

The Chinese also had a mid-range calendar with a cycle of 60 years, that was based on the reoccurring conjunction of Jupiter with respect to certain positions in the sky. Whether these cycles served to reconcile their lunar short calendars with the fixed stars in the sky, or something more mythical, is not known.

The Aztecs also had a 365 day long solar calendar, clearly observant of the Sun cycles. There is an original artifact at the National Museum of Anthropology in Mexico City, the famous Aztecs Sun Stone. The calendar appears to have been started with a day referencing a special celestial event, the spring or vernal equinox. That is the day when the length of the day and the night are the same, although we do not know how they measured the duration of the day and night.

The best reconciliation of the lunar-solar discrepancy is due to the Babylonians, also several millennia ago. Their long term observation led to the recognition of an intriguing coincidence. They recognized that a lunar cycle or lunar month is about 29.5 days. They noticed that 19 solar years is exactly 235 lunar months. On the other hand that is 19 lunar years plus 7 lunar months ($19 \cdot 12 + 7 = 235$). Hence one could live with a lunar calendar for 19 years and then add 7 lunar months to get back in synchrony with the skies.

That approach, however, would have made them out of synchrony with the seasons for an extended amount of time. Hence they decided to introduce the additional lunar months in installments, a technique that became instrumental in the calendar making and correcting efforts for millennia after that, and even today. They attempted to distribute the 7 months through the 19 years in monthly installments and in regular intervals as much as it was possible.

In a solution reminiscent of our current leap year technique, they added an extra month at the end of the 3rd, 6th, 8th, 11th, 14th, 16th and 19th year of the 19 year cycle. This rather insightful solution resulted in a calendar that was never more than 20 days out of synchrony with the solar cycle. Hence this calendar was also a luni-solar calendar based on lunar months, but attempting to stay in synchrony with the Sun.

The notable role of the number 7 prompted the Babylonians to consider it unlucky. It is believed that the appearance of the week as a time duration was invented by the Babylonians when they declared the seventh day to be the day of rest to avoid any unfortunate events. It is assumed to be the origin of the biblical statements of creation in the book of Genesis: God working on the creation for six days and resting on the seventh. Even in our modern calendars the order of the days is Monday through Sunday, the work week starts on Monday and the rest day is Sunday.

The Babylonian calendar became very popular in the Mediterranean region in antiquity. Most of the surrounding cultures adopted it. The shortcoming of the calendar was, however, the adjustment years. Their timely execution was dependent on a contiguous rule or governance, and of course an ongoing record keeping. Considering the volatile nature of the region, the rulers and changing allegiances, it was liable to fall out of synchrony.

As we now know and see on the next figure, Earth is orbiting on an elliptical pattern, and the Sun is in one of the focal points of the orbit. Hence when the Earth is on the side of its orbit closer to the Sun, the so-called perihelion, it is moving faster than on the opposite side, the aphelion.

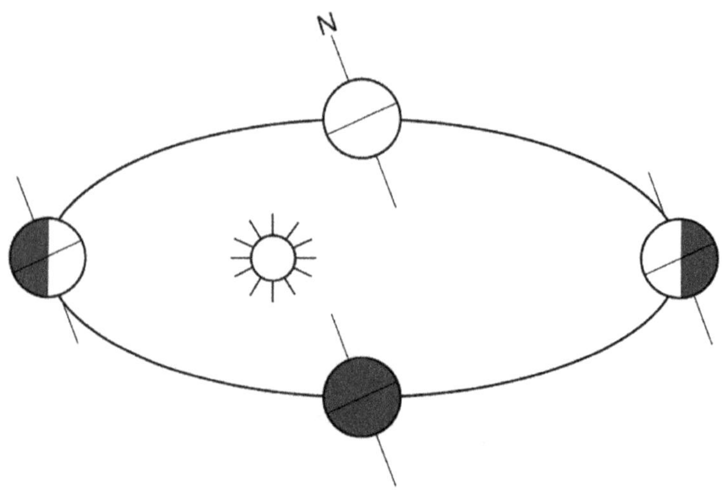

A solar year

The location of Earth on the left side of the figure above represents the winter solstice of December 21st, a time of fear in ancient cultures. It followed a continuing lowering of the Sun's height day by day and people feared that one day the process would not stop and Sun would not come up at all.

Then of course things would turn around and spring would come with lengthening days and higher locations of the Sun. The bottom location on the figure is the spring equinox of March 20/21st. This is the day of equal length in days and nights.

The Sun continues to be higher and the days longer until the summer solstice on June 21st. This is the longest daylight day of the year and its position is on the right hand side of the figure. This was a joyous event in many cultures, the so-called mid-summer day festivities are still preserved.

Finally, the cycle concludes by gradually lower Sun positions all the way until the fall equinox on September 21/22nd. It is again the date of equal length of day and night. The lowering heights and shortening days continue until the winter solstice. Apologies to the readers on the Southern hemisphere, for whom the above explanations should be reversed.

Considering the line representing Earth's axis of rotation pointing to north, and comparing the winter and summer solstices, we can see the different sunlight patterns on the two hemispheres. The line representing the equator shows how during the winter in the northern hemisphere, the sun is below, and during the summer above the equator.

While the figure implies that Earth is always leaning in one direction, that is far from the truth. The axis of Earth as shown on the figure is not perpendicular to the orbital plane of the Earth. This is in part due to the gravitational forces of the Sun and the Moon. The forces attempt to pull the equatorial bulge of Earth into the plane of Earth's orbit, the plane of the ecliptic. The axis is slowly rotating in the so-called axial precession, that is is a rather large, almost 25,760 year long cycle.

During the precession cycle the axis of Earth is going to change such that Earth's Northern hemisphere is going to be having winter on the right hand side, i.e. the farther away from Sun. Whether this time will be accompanied by a reversal of the magnetic field of Earth is subject to much debate. Let it be sufficient to state that there is a much bigger celestial cycle than the yearly solar cycle.

If Earth would have a perfectly circular orbit around the Sun and its axis would have no tilt, Sun would be always at the same point in the sky at the same hour of every day. Since neither of those conditions exist, the location of the Sun in the sky varies from day to day. Recording the location over a year yields a very intriguing pattern of a figure 8. The pattern is called the analemma is shown on the next figure. The pattern was probably recognized in the antiquities.

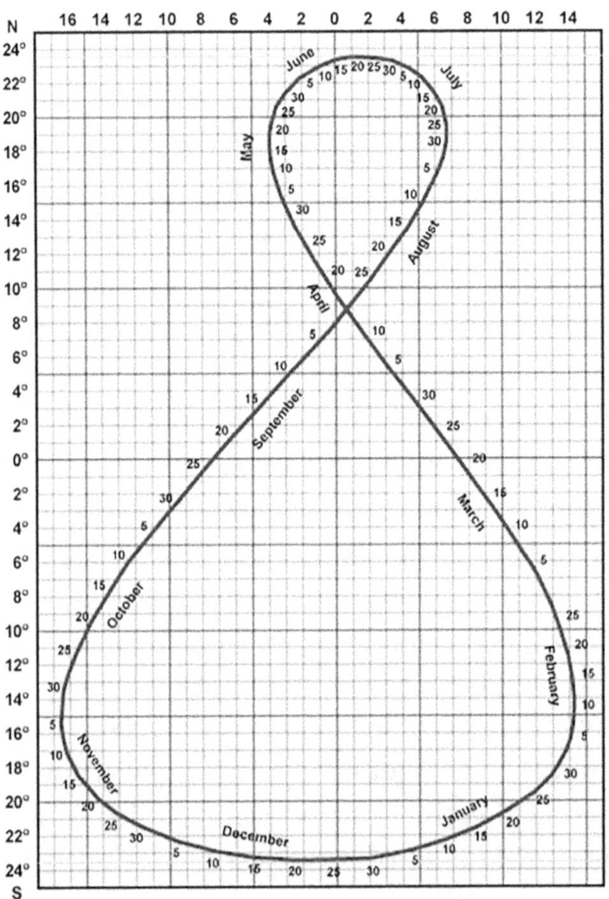

The analemma

One only needs a fixed position retained year around and an unchanging orientation toward the sky to observe the pattern. Ancient astronomers probably had both conditions. It is actually possible to photograph the pattern by modern equipment on a single piece of film repeatedly exposing it on every day in the year. The camera must be permanently mounted and the pictures taken at the same time on the different dates.

More importantly, the pattern is a very useful time keeping tool. Its vertical axis gives the latitude at which the Sun is directly overhead on a certain day. Its upper maximum is at 23.5 degrees North, which is the tilt of the Earth's axis and the lower minimum is at 23.5 degrees South. The dates are June 21 and December 21, the days of the summer and winter solstices. The latitude locations are the Tropic of Cancer and Capricorn, respectively.

The horizontal axis represents the time difference between the Sun's position and the local time. On the right hand side of the vertical axis the Sun is behind the local time (slower) and on the left hand side it is ahead of the local time (faster). This is due to the fact that the length of the solar day is not constant during the year. When the Earth is on the side closer to the Sun it moves faster, the solar day is shorter, by as much as 16 minutes. On the other side it may be longer by about the same amount. The symmetry of the arrangement results in the average value of $24 \cdot 60 = 1440$ minutes a day.

The analemma in essence is a graphical representation of the seasons. Changing seasons were of utmost importance to the Egyptians, a culture highly dependent on the yearly floods of the Nile. They opened their eyes toward the stars in order to produce a repeatable yearly cycle. About five millennia ago they recognized the very regular behavior of a bright star appearing on the early morning horizon, just before sunrise. The star, we now call Sirius, appeared very close to the Sun in early September, our current time, just preceding the yearly flood of Nile. Hence it was called the flood bringer.

Their yearly calendar start was aligned in close proximity with the rising of Sirius and completely abandoned the lunar component of the calendar. They defined artificial but fixed length months of 30 days each and 12 months of those constituted a year. This was only 360 days, just like the Mayans', hence they also introduced a 13th month with five days at the end of each calendar year.

This calendar was installed by the pharaoh Imhotep, sometimes in the 27th century BC. The year had 3 seasons of four months each. The season of the flood was from September through December, according to our current system. The next four months were the season of the vegetation, from January through April. The months from May through August represented the season of the harvest. The calendar was still slightly out of synchrony with the solar cycle, since it was 365 days long exactly.

In 238 BC, Ptolemy III decreed that every fourth year should be 366 days, resulting in a 365.25 days long average over 4 years. The adjustment schedule was kept by the high priests of the church and became known as the Ptolemaic calendar.

So why is the solar year not exactly 360 days long? After all, one full rotation of the Earth around the Sun is 360 degrees, and one full rotation of the Earth around its own axis is one day. The reason is that Earth simultaneously rotates around its own axis and around the Sun.

Completing one full rotation around its own axis, Earth does not arrive in the same position with respect to the Sun. Each day Earth has to compensate for the fact of being on the orbit around the Sun and rotate a bit more to show the exact same face to the Sun. That is the reason why the number of Earth rotations in a year is not exactly 360.

This argument is true independently of the units of measuring the angle. If Earth were to spin around its own axis faster but on the same orbit around Sun, the number of days in a year would still not be a whole number. That the measure of angle for a full circular motion turned out to be 360 degrees in human history, is simply rooted in the fact that Earth takes approximately that many rotations in a year.

16

LUNAR MONTHS

Since the Roman empire's territories spread from the deep south in Egypt to far north all the way to the British isles, the agreement between the calendar and the seasonal changes of the Sun was very important. The daylight hours at various latitudes and certain times of the year could vary a lot between the extremities of the empire.

People living close to the Equator and even in Egypt do not see much of a variation of daylight hours during the year, but this is not the case for people living on the northern latitudes. For example, above the 50th latitude, like today's London, in the winter months of the year the daylight hours maybe as short as 10 hours or less. On the other hand, during the summer months daylight hours could approach 18 hours.

The ancient Roman calendar, assumedly established by Romulus, the founder of Rome had only ten months originally. The months were: Martius, Aprilis, Maius, Iunius, Quintilis, Sextilis, September, October, November, December. It is noticeable that some months follow the Latin name of their numeral order, while some do not. Quintilis, Sextilis, September, October, November and December verbally mean the 5th, 6th,

7th, 8th, 9th and the 10th months. But they are not anymore and we will soon see why.

Martius was named for the God of war, Maius and Iunius were for Roman Goddesses. The meaning of Aprilis may have been designated to honor Aphrodite, the Greek name of Venus. The ten months long calendar with 29 or 30 day months, still trying to follow the lunar cycles, was of course extremely short, less than 300 days. One of the later Roman rulers in the 7th or 8th Century BC added two additional months, Ianuarius and Februarius to the end of the year and the calendar became 355 days long.

With such a year, the Romans had a noticeable difference when they compared the calendar day dictating the spring or autumn equinox and its day based on, by that time very accurate measurements. Hence some adjustment was needed.

This was accomplished by a convoluted rule of inserting days into certain years out of every 24 years. These insertions were 22 days each and the insertion years became 377 days long. This rule resulted in an average year of 365 days over a 24 year period. This was clearly an improvement, but it was a system of high complexity.

Therefore it was managed by the Roman magistrate, who dictated the length of each year according to the general principle above. This fact, however, allowed political tampering with the system. The chief magistrate could lengthen the year in which sympathetic politicians were in office and shorten the year of the opponents. This happened on several occasions, sometimes to the extent of a complete misalignment with the solar year. Roman history reports several so-called years of confusion.

Julius Caesar was born in 100 BC (according to our current time counting). He was born into a noble family; his father reached the office of praetor which was the second highest public office. Julius himself engaged in public service early on and also reached the office of praetor by 62 BC. Two years later, in 60 BC he was elected to consul, the highest office. A decade of military activity followed and culminated in his famous crossing of the Rubicon river in 49 BC to end the ongoing civil war in Rome. As a result Julius Caesar became dictator in 48 BC.

By the time of Julius Caesar, the Roman calendar was seriously out of date. Several insertions were missed and celestial events of solstices and equinoxes were falling onto the wrong days or even into the wrong months. Julius Caesar set out to correct the issue, once and forever. While his goal was not fully achieved, his effort was nevertheless very noble, resulting in a system that survived another millennium and a half.

Despite his active involvement in the affairs of the Roman state, Caesar continued his military activities in the two years following his installation as dictator. He successfully conquered Egypt and had a famous love affair with Cleopatra, the Egyptian queen. This, however, may have been more than just a romance of a middle aged man with a young beauty, many years his junior. He became very appreciative of the Egyptian culture whose accomplishments were brought to his attention by the superior intellect of Cleopatra.

It is likely that during that time Caesar became aware of the calendar work of Egyptian astronomers, who had been observing the celestial cycles for thousands of years. They compiled their observations of solstices and equinoxes in detailed records. Caesar supposedly received the method of repairing the Roman calendar from an Egyptian astronomer, named Sosigenes of Alexandria. His recommendation to Caesar was the rather ambitious idea of abolishing the irregular insertion of days and years as they were before, and replacing them with a new system that started with a major adjustment and followed by regular minor insertions.

Caesar apparently followed the advice and issued his reformed calendar sometimes during the summer of 46 BC, in the month then named Quintilis, the fifth month. He needed to add 67 days to the calendar. The 67 days were necessary because of the missing insertions in prior years. To accommodate the extra 67 days, he cleverly moved the by then last two months following December, that were January and February, to become the first two months, thereby shifting the others.

Another way to view this is that he just restated the start of year from March 1st, to January 1st. After all, the year is only a cycle whose beginning and end could be arbitrarily chosen. He may have learned that from the Egyptian calendar whose beginning was in September. Either way, this put his date in the month of Quintilis in proper alignment with the solar year.

But this was not enough for 67 days, so he added to the lengths of the various months, hence some became 30, some 31 days long:

31, 30, 31, 30, 31, 30, 31, 30, 31, 30, 31, 30.

His order was a regular alternating sequence of long and short months, but now the year was 366 days long. In order to correct this, he proposed that February be 30 days long only once every

4th year. The other three years it was to be 29 days, resulting in a 365 and a quarter day average over every four year period, just like in the Ptolemaic calendar. The fourth year, when February is longer, is what we call the leap year today.

This naming convention is worthy of an explanation. What happens in that year is that if February 28th is a Monday in a leap year, the same date of February 28th will be a Wednesday in the following year. This is due to the extra day that results in the Tuesday being "leaped over" in the next year. A simple rule that cannot be ruined, so it seemed.

There was a mistake, however, in its implementation. The extra day was added every three years, instead of every fourth, and the calendar got out of synchrony soon again. The reason for the mistake is unclear, but records show that 45 BC was declared a leap year, possibly to account for the missed 46, that was supposed to be the first leap year. The next legitimate leap year was 42 BC and this three years difference created an unfortunate precedent. The leap date insertion continued every third year all the way until 8 BC.

The Senate later honored Julius Caesar by renaming the original 5th month, Quintilis to Julius, in appreciation of his contribution to the Republic. Julius was assassinated on March 15, 44 by his adopted son, Brutus. His death resulted in a huge societal turmoil that lasted more than a decade. The state of Roman affairs did not really stabilize until Augustus, the first ruler of the new Roman Empire, as opposed to Republic. Augustus, a nephew of Julius Caesar ruled from 27 BC to 14 AD, thus transcending the time of the birth of Jesus.

He continued Julius' work on the calendar. It is notable that two relatives contributed to the correction of the Roman calendar. He corrected the leap year spacing error in 8 BC, abolished the three year leap cycle and prescribed it to be four.

According to historical records, Augustus himself changed the eight month, originally called Sextilis, to Augustus, in a bit of a self-promotion. Some other rulers with monumental egos, also made attempts at renaming certain months, but they did not stick. Nero, for instance decreed the new name of Aprilis as Neroneus, but after his bloody reign, nobody wanted to remember him every year.

Furthermore, some anecdotal records indicate that Augustus detested the fact that Julius was 31 days and Augustus only 30. Therefore he "stole" one day from February to make it the 28/29 days as we know it today. He added it to Augustus, to make it also 31. Then in order to still maintain a long/short sequence, he changed the rest accordingly to:

31, (28/29), 31, 30, 31, 30, 31, 31, 30, 31, 30, 31.

This irregular long/short sequence of calendar days still causes problems to countless generations of school children having to learn the sequence of the months by engaging in the hand game of counting knuckles and valleys.

The most common method of counting the years by the Romans was by counting them from the inauguration of the new pair of consuls. Beginning in 153 BC (per our present counting) they usually started the office on New Year's day. The years counted in this fashion were also sometimes called the consular years. This counting was cyclical, the year count was set back to zero at the beginning of every new consular period. Well, maybe to one, the zero as an algebraic entity was not widely accepted in Europe yet.

There was yet another system, in which the years were counted from the founding of Rome. This was, however, not used by the common folk in everyday circumstances, mainly because its backward time horizon was just excessively long (although such a fact did not bother the ancient Mayans at all). This was used mainly for official record keeping by ancient historians.

To make the interpretation of the Roman calendar record even more convoluted, years were sometimes counted starting from the beginning of the reign of an emperor. The years in this counting system were called the regal years.

Both consular and regal year counting were in use all the way past the middle of the first millennium. Their usage attests to the difficulties of humankind to establish a uniform time scale even within one particular empire, let alone world-wide.

Finally, in 525 AD (per our current count) the philosopher Dionysius Exiquus proposed to align the year counting to the birth date of Jesus. His system of Anno Domini (AD or the year of the Lord) soon permeated the whole Christian world. He denoted the dates before the first year AD by BC for Before Christ.

There are conflicting stories about the correctness of the birth date. According to one school of thought, he was referring to the "incarnation of Jesus' that would put it to the date of

Jesus' conception, not actual birth, but the date is by and large OK. There is another camp of historians who propose that Dionysius actually made an error of several years.

They base their hypothesis on the motion of Jupiter, considered to be the factual star of Bethlehem. Most of the time the outer planets, such as Jupiter appear to be moving west to east, since we on the Earth are on the inner circle. But at the proximity of the point when Earth "overtakes" an outer planet, the phenomenon of retrograde motion occurs. During this, the outer planet appears to be moving opposite to the normal direction, i.e. from east to west.

According to astronomical retro-calculations, starting from August 23rd until December 19th in 6 BC per our current count, Jupiter was in its retrograde motion. This could have been the star the three kings followed west to Bethlehem according the Bible.

Another historical fact supporting a date other than the later chosen AD 1 is that according to some biblical records Jesus was born under the reign of Herod the Great. Herod, however, died in 4 BC, a well documented fact. It appears likely that Dionysius was off by about 6 years, but it does not seem to matter for our current state of calendar keeping.

Even with the adjustment day and the leap year sequence, the Roman calendar accumulated noticeable error through the first millennium. The Julian calendar year (in agreement with its origin, the Ptolemaic) was exactly 365 days and 6 hours long. The solar year, as we now know very accurately is 365 days, 5 hours, 48 minutes and 46 seconds. The discrepancy is 11 minutes and 14 seconds, or 674 seconds. Since a day is 86,400 seconds, the 674 seconds yearly error of the Julian calendar amounts to a full day error in every 128.19 years.

While that does not seem to be of big consequence, it is significant in historical perspective. The Julian calendar fell into considerable discrepancy regarding certain calendar days of societal importance, like the beginning of Easter. That date was originally defined in a specific alignment with the spring equinox of March 21. By the middle of the second millennium, the Julian calendar was about ten days ahead of the celestial schedule and this required another correction.

Pope Gregory XIII was born in 1502 as Ugo Boncompagni and died in 1585. He studied law in his youth and started a legal career that included teaching law to some of the high members of the clergy. He lived a lay early life, even had a son out of marriage, but later was ordained. By his late thirties he was working for various papal offices, including being a legate to Spain. He became a cardinal in 1565 and was elected to Papacy in 1572.

His legal background likely contributed to his reform minded reign as Pope; he set out to introduce significant changes to the Catholic church. Those included positive rules that required that bishops reside in the district they oversee, and some negative ones such as issuing a list of forbidden books.

His most important contribution as a reformer was, however, that he organized a committee to work on mending the Julian calendar. His interest was naturally closely related to the church, specifically to that of the date of Easter.

The committee included several disagreeing members, among them Aloysius Lilius, a medical doctor, whose idea prevailed in the end. He proposed to use a length of the year published in the middle of the 13th century, in the so-called Alfonsine tables. They were named after the Spanish king Alfonse X, who commissioned their publication.

The Alfonsine tables listed a mean calendar year of 365 days, 5 hours, 49 minutes and 16 seconds. That was still 30 seconds longer than the true calendar year of 365 days, 5 hours, 48 minutes and 46 seconds. This was, however, much better than the Julian difference of 674 seconds.

Lilius' main contribution was to find the way to bring the Julian calendar into synchrony; a difficult thing to accomplish. The difference between the two was 644 seconds. Dividing 86,400, the number of seconds in a day by 644 yielded 134.16. This meant that the Julian and Alfonsine count was off by a day in every 134 years.

Some other members of the committee proposed simply dropping one day in every 134 years, certainly a conceptually simple, but administratively cumbersome if not impossible solution. Lilius, however, recognized that $3 \cdot 134 = 402$, or almost 400 years. This led to the easily manageable rule of ignoring three leap years in every 400 years which is still the basis of the current leap year rule.

Going back to Gregory's main concern, the date of Easter was a confusing topic for hundreds of years and it is somewhat clouded even today. The various fragments of the Catholic church celebrated Easter on different days. The Catholic church at the time of Gregory defined Easter as the Sunday after the first full moon that follows the spring equinox. The date of the spring equinox was March 21st, hence the definition could place Easter anywhere between March 22nd and April 25th.

Due to Easter's religious importance, establishing the precise date of the spring equinox was crucial in the calendar. By 1582, the accumulated error in the Julian calendar showed March 11 as the day of the spring equinox that was supposed to be March 21. Clearly there was a discrepancy that needed to be corrected.

The Papal decree by Gregory in 1582 ordered the day after October 4 not to be October 5, but the 15th. This change, however, only addressed the discrepancy between the Julian calendar and the celestial cycles. Without any further change the new calendar would also gradually become misaligned in the future.

Hence the reform also required the modification of the Julian leap years, as proposed by Lilius. The redefined rule was that the leap years (that are in general in each fourth year) are skipped when the year is a full hundred but not divisible by 400. Hence 2000 was a leap year, while 1900 was not and 2100 will not be.

With this new definition, the average Julian year of 365.25 days was reduced to 365.2425 days. Precisely, the Gregorian year became 365 days, 5 hours, 49 minutes and 12 seconds long. The deviation from the solar year became only 26 seconds. This, still approximate year length decreased the accumulated discrepancy by several orders of magnitude. No adjustment is going to be required for about a thousand calendar years.

17

HOURLY CYCLES

The Egyptians had sun-dial devices as early as 1500 BC. They were sticks stuck into the ground and the time measurement was done by observing the shadow of the stick. The stick cast a long shadow at sunrise, toward west. As the Sun was rising, the shadow came closer to the stick, until by noon it was close to the foot of the stick. On certain locations, like on the Tropic of Cancer the shadow actually disappeared at noon, when the Sun was exactly overhead. Then the afternoon Sun turned the shadow toward east and gradually increased its length. By sunset, the long shadow was pointing to the east, reversing the progress from the morning.

While it was possible to rely on the Sun to measure the passage of time during the day via sundials, the continuous nature of Sun's motion made the quantitative measurement difficult. Where was 9 am, or 3 pm on the sundial shadow? This begs the question: why is the day 24 hours long? According to the current wisdom, the use of 12 over 10 was due the Egyptians, who were very adept in counting fractions and working with them in various algebraic computations. Their preference of 12 over 10 may have been because 12 has more integer factors: 2, 3, 4, and 6, while 10 has only two: 2 and 5.

Their algebraic preference may also have been reinforced by celestial events. Some records indicate that in ancient Egypt in the summer 12 stars arose during the night. Their rising was, of course, not exactly hourly per our current clocks, but their number was the defining fact. In any case, the hourly timekeeping became standard only with the advance of mechanical clocks, developed much later.

The sundials also had a problem on overcast days and during the night, due to lack of sunshine. The next advancement in time measuring devices were the water clocks, also invented by the ancient Egyptians. Their big advantage over the sundials was their ability to keep time during the night also. Their disadvantage was the need to regularly replenish the water.

To alleviate the need for the regular water refill, the water clocks were later connected with a continuous water source. To quantify the now continuous time flow, various mechanical instruments were invented. The most common versions relied on a floating device that had some stylus attached. The stylus pointed to a scale with markings for certain time durations, like hours. The continuously flowing water filled a barrel and the rising water level lifted the flotation device, most likely cork. It was still necessary to reset the clock by emptying the barrel when it became full, but only once a day.

Water devices were not portable, however, therefore self-contained sand based hourglass timekeeping devices dominated the middle ages. This was especially true to the most common means of travel at the time, ships. Some records indicate that Ferdinand Magellan, during his historic circumnavigation of the Earth, used 18 hourglasses to measure the time of daylight. The first glass was turned on at sunrise, then when it wound down, the second was turned upside down and so on. The number of wound down glasses showed the number of hours elapsed since sunrise.

Observing repetitive rotational phenomena in the sky was a basis for time keeping since the antiquities. No surprise then, that combining the gear concept with a little of Newton's gravity resulted in the weight operated clock. It found employ in the clock towers of medieval cities or in the still popular grandfather clocks.

Their mechanism is simple: a cylinder is rotated by a winding key to roll up a string holding a weight onto the cylinder. The gradual pull of the gravity on the weight will unroll the string from the cylinder and the gear mechanism on the axis of the cylinder will convert that motion into the clock's progression. Two different gears operate the distinct hour and minute arms of the clock at different speeds. Some also have additional time scales, like Moon dials showing the phases of the Moon, requiring a delicate set of gears inside.

Some of the weight operated clocks also contain a chime, a rather loud one in case of the clock towers of towns. This is accomplished by another weight, usually larger than the one keeping the time. Then there are the cuckoo clocks having a birdie pop out once in a while and even sing a melody. Another weight might be used for that purpose. The size of the weight is usually dictated by the torque necessary for the particular operation. The elaborate Glockenspiels in the towers of European towns are driven by very heavy weights suspended on rather long wires to be able to sustain the lengthy dance motion of the characters, a huge tourist attraction in some cities.

The idea of a pendulum clock was originated by Galileo Galilei, a frequent hero of our story. Anecdotal evidences state that he recognized the uniform motion of a suspended pendulum and the independence of the frequency of motion from the mass of the pendulum. Apparently he realized this fact while watching the lanterns hanging from the ceiling of his church. He actually designed a clock based on a pendulum, but he was not able to find a craftsman to produce it.

Christian Huygens, the famous Dutch physicist completed Galileo's work and created the first working pendulum clock. This enabled the measurement of fixed seconds and Huygens defined the 3,600 seconds long hour duration. The origins of the 60 base for time keeping is attributed to Babylonian place system of the same base. Huygens' original pendulum clock, built by a Leiden clock maker in the 1650s is still exhibited in a museum there.

Huygens was born at The Hague in 1629 and also died in his city of birth in 1695 after a spectacularly successful scientific life. He studied at the University of Leiden, still a scientific stronghold today. His main interest was physics and his scientific accomplishments are impressive. His theory of the light waves is his most important scientific legacy.

He also noted that the pendulum's period is a function of the location of the weight, i.e. the pendulum's length, and not its mass. This was important, because this enabled the adjustment of the clock. Still today, the weight at the end of the pendulum (the bob) is on a thread, hence the length of the pendulum and its accuracy may be adjusted by synchronizing against another clock.

Furthermore, he recognized the fact that the pendulum's swings are not isochronous, or executing in the same time duration. There is certain decay in the period of the pendulum swing that depends on the hinge mechanism and its friction. Huygens devised a mechanism that advanced a gear at every swing and at the same time exerted a nudging force on the pendulum to retain its cycle time. This was the final component to make the pendulum a reliable time measuring device, truly a clock.

It was known to the early Mediterranean sailors millennia ago that the longitude location is directly related to time. Since the Earth rotates 360 degrees in 24 hours, it is a simple division to figure out that it rotates 15 degrees per hour. As a consequence, when a ship is traveling on the sea, for every 15 degree change in longitude the local time varies by an hour. The ship traveling east looses one hour of the local time, i.e. the clock moves ahead. This is manifested today, when one is taking an airplane to the east. The day is very short when flying non-stop from California to Europe. Leaving in the morning gets the plane into Europe the following morning. The explanation is the 9 or 10 hours time difference due to the time zones plus the 10-12 hours flying time.

Conversely, a plane traveling west benefits from the direction and gains an hour in every 15 degrees traveled westward. Leaving Europe in the morning puts us into the California by early afternoon, despite the fact that the flight time is 10-12 hours. This gets a bit more confusing when one travels between the US and Asia. Strange things happen when one crosses the international date line, leading to the longitude problem.

The problem is the difficulty of keeping a reference time when traveling on the open ocean on ships. While the local time was measurable based on the Sun, in order to find out the longitude location the time at a fixed time point was also needed. Simply a device was needed on board that was able to reliably keep the time at the home port. Unfortunately neither the weight, nor the pendulum clocks were useful on ships where they were not in vertical position.

In the 17th century, prior to time zones and airplane travel, the problem was difficult. King Charles II in 1675 issued a challenge to the scientists at the Royal Observatory to solve it. Despite almost a half a century of research and numerous failed solutions, the problem remained unresolved. The British admiralty offered a reward of 20,000 pounds sterling in 1715 to solve the problem with an accuracy of half a degree in longitude, or two minutes of time.

John Harrison was born in Yorkshire in 1693 and followed the footsteps of his father by becoming a carpenter. He received no formal science education but still, became very interested in time keeping mechanisms already as a teenager. He built a wooden clock true to his trade in his early 20s and developed a reputable clock business with his younger brother, strictly using wood in his clocks.

Later he became engrossed with the longitude problem. The problem needed a time keeping device that was portable and not subject to the physical effects acting on the ship in the seas. He built a portable version of his wooden clocks in the early 1730s with moving parts that were controlled by springs and screws, hence completely independent of gravity. This was an important consideration due to the motion of the ships.

Harrison's chronometer was tested on a voyage to Lisbon aboard the ship named Centurion. It performed very well, however, was deemed not to reach the accuracy requirement of the Admiralty. This evaluation is highly suspect in retrospect, after all, how did they measure a more accurate solution when there was no such device available?

Nevertheless, Harrison took on the challenge of further improving his device. He spent almost three more decades and redesigned his device three times. Another test was sanctioned by the Admiralty in 1764 and this time Harrison's son, William accompanied the clock on a trip to Barbados. The fourth incarnation of Harrison's design was on board and was accurate to an error of less than 40 seconds over a journey of 47 days. By all, it was considered a great success, but for the Admiralty.

They declared the clock to be a lucky, one of a kind piece of work and refused to issue the price. They demanded from Harrison to turn over the design to the Admiralty for 10,000 pound sterling. The second half of the price would be paid only when other devices built based on the design proved to be accurate on 30 mile long longitudinal time trials. This was in 1765.

What would follow is a long story, worthy of another book on its own right. It shows the pettiness of the establishment regarding the commoner who answered their challenge. Despite several copies made, independently tested and proven to be three times as accurate as the requirement, the Admiralty still refused to pay off the second half of the price. On the other hand the elderly Harrison, in his 70s by that time, still refused to give up.

An interesting cycle of life concludes the story. In 1772 the feud got to the new king, George III, whose predecessor originated the challenge. He himself evaluated the clock and was impressed with the results. Still, the Admiralty did not pay. It took an act of Parliament in 1773 to force a second payment. Harrison was vindicated.

Harrison was even further acknowledged by Captain Cook, who took one of Harrison's clocks with him on his famed three years long trip. Upon returning he announced that the clock never erred more than 8 seconds during the whole voyage and declared it to be their "faithful guide". The challenge was answered, the ability to measure longitude by a clock was achieved. Harrison died in his 80s, a year after Cook's return, likely a satisfied and proud person.

It is fitting that the final chapter of the longitude challenge, that was instigated by the early years of sea travel toward the New World, starts with the contribution from the New World. Besides a tool to keep the reference time, there was a need for a common reference location. This was established in Washington D.C. in 1884 on the so-called Washington Meridian Conference, a gathering of the most important maritime nations.

The participants in the conference agreed to designate the meridian (longitude line) passing through Greenwich, now a suburb of London, to be the Prime Meridian, or the zero longitude line. The degrees of longitudes were to be measured to the east and west, up to 180 degrees. The 180 degree longitude (either east or west) became the international dateline. Using the dual possibility of measuring the longitude in degrees or in time, the prime meridian at Greenwich is 0 hours zero minutes and the international dateline marks 12 hours and zero minutes.

The international dateline is sometimes also known as the Sunday/Monday line. The line is running down in the middle of the Pacific and when it is Sunday 12:00 noon on its eastern side, on its western side it is Monday at the same hour. The line is occasionally moved a few degrees for small distances to accommodate some of the island nations straddling the line. After all, it would be rather confusing to have some villages with different days on two sides of a street.

The time measurement of the longitude led to the time zones. Every 15 degrees in longitude, as we have shown earlier, amounts to an hour change in local time. For example, the location of the writing of this book, California is at 118.24 degrees west from Greenwich, or the time here is 8 hours and almost 5 minutes behind Greenwich. When the Greenwich schools open at 8 am, the California school children are sleeping at midnight.

The cities within an hour time zone of 15 degrees of longitudinal range all adhere to the same time. The time zones are in hourly increments, except again on the Pacific, where in some strange cases half an hour time zones exist. Finally, some nations, like the United States, adjust the time zones seasonally, leading to Daylight Saving Time. This topic is interesting when some states in the nation do not adhere to the rule and result in some confusion, but we digress.

Measuring local time at the poles is somewhat tricky because all the longitudinal lines coalesce into a point. It appears that while standing on the pole, we can align ourselves with any time zone we like. For reasons of consistency, sometimes the North Pole is assigned to meridian zero, or Greenwich, while the South Pole is assigned to the international date line.

Humankind continued to invent new clocks of various kinds. The name of the device probably comes from the Latin word clocca, meaning bell. Since on ships the passing of time was usually marked by ringing a bell every time an hourglass wound down, this led to the name. Then came the more advanced devices in church spires and the accompanying bell ring at every hour, or later even a chime at the quarters. Those were driven by gravity via weights attached to a rotating cylinder as they still are today.

In the next advancement of the clock technology the gravity was replaced by winding up a torsional spring. Still rotation ruled, even when the spring was later replaced by the battery. We seem to have overcome the need for rotation in measuring time only with the appearance of the digital clocks, although some of those also have analog faces with rotating arms. And there are still many, who regularly wind up their wrist watches inherited from their grandfathers.

Swiss watchmakers are credited with creating the first commercially available spring watches. These were also the first portable watches. At the beginning people were wearing them on a chain around their neck and it was a rather prestigious symbol of status to own one. They were also serving as jewelry. Diamond or gem stones were embedded into the face that was not yet covered by glass as we know it today. There was a cover, sometimes solid on a hinge, and sometimes a grill work enabling to see the time without opening it. It is also believed that the originally square portable watches evolved into a round shape to ease removing them from the pocket.

The second half of the 16th century brought the minute hand by Swiss clock makers, whose mastery of course is still admired today. Some records indicate that the very first clock with a minute hand was built for Tycho Brahe, the famous Danish astronomer, to aid his celestial observations.

Humans felt that the heart beats many times during a minute and considered this to be a reason to create a smaller unit to measure. That unit became the second. The origin of the second as the smaller time unit is actually the remnant of the original designation of "second minute".

18

ROTARY MACHINERY

Since we arrived at a common rotational equipment in the clocks and watches, we can follow this avenue further into our everyday life. Once the gear concept was known, humankind was able to do wonders with the rotational motion. With a properly chosen tooth pairing, the gradation of the rotational speeds became possible.

This is the concept utilized in the bicycle where the rotational motion provided by human pedaling is geared up or down to turn the wheels on the ground slower or faster depending on the terrain. The gear is of course carried much further in our automobiles and other vehicular structures operating under machine power.

After being able to keep time, it was time to put the rotational phenomenon truly to work. Perhaps the most important rotary machinery of our lives is the electromagnetic kind. It all started with a Danish physicist, Hans Christian Orsted, who around 1820 demonstrated that a simple magnetic compass is sensitive to an electrical current. The electrical phenomenon itself was in its infancy at the time and Orsted's observation connecting it with the magnetic phenomenon opened up a wonderful world.

Orsted was born in 1777 in Rudkobing in Denmark. His father was a pharmacist and the young Orsted was fascinated with chemistry while growing up. By 1799 he earned his doctorate at the University of Copenhagen and set out to spend several years of postdoctoral studies in Germany where his interest in physics deepened. Upon returning to his alma mater in Copenhagen, he became a professor and spent the following decades in research and teaching.

It appears that his serendipitous discovery of the phenomenon occurred while preparing for an experiment in his lecture with a compass laying on the table. Orsted noticed that when he placed the compass in the proximity of a simple electric circuit powered by a battery turned on, the compass moved. When the battery was turned off the compass regained its natural position, aligned with the Earth's magnetic field and pointed north. It appeared that the electricity flowing through the wire created a magnetic field of its own overriding that of the Earth's.

A few years later, Michael Faraday, an English physicist conjectured that the reverse phenomenon must also exist. Faraday was born about a quarter of a century after Orsted in a London suburb, into a poor family of several children. Hence, after completing his grade school at 14, he was put into a bookbinder apprenticeship to learn a trade. The apprenticeship lasted seven years, during which the young Faraday, eager to learn, read many books on chemistry and physics. By the time he completed his apprenticeship, he had learned enough to be able to follow lectures at the Royal Society using free visitor tickets given away by a charity organization. It was surely an unusual training for a scientist.

Serendipity played a role in his life as well when he obtained the role of an assistant to the known chemist Humphry Davy of the safety lamp fame, one of the distinguished scientists whose lectures he attended. This provided an opportunity for him to access a laboratory where he was able to devise and execute his own experiments.

One of those led to his discovery that a moving magnetic field results in an electric current. His experiment contained a simple closed electric circle without a battery, but with a measuring device. He used a strong permanent magnet and moved it around the wire. Sure enough, the device showed current flowing through the circuit.

He connected that to Orsted's result of a moving electric field producing a magnetic field (moving the compass) and created the since inseparable pair of electromagnetism. That became the basis of many rotary equipments of our life.

A most important pair, in true correspondence to the principle, is the electric motor and the electric generator. The electric motor produces mechanical energy by a changing electric field, somewhat similarly to Orsted's experiment. The following figure depicts the concept of a simple electric motor. The wiring surrounds two electric magnets on the perimeter and a permanent magnet in the center constitutes the rotating part. In practical motors there are many more of these components, but the concept is the same.

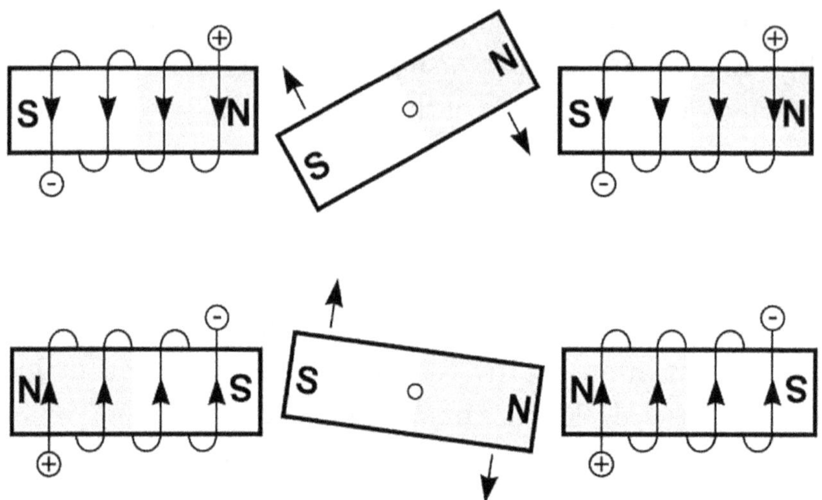

The electric motor concept

Systematic change of the electric current in the external (stator) part induces a rotating motion in the internal (rotor) part of the motor. Specifically, the changing current in the stator circuits creates an alternating magnetic field. The field is either pulling the rotor magnet as shown on the upper part of the figure, or pushing as shown on the lower part, depending on the magnetic pole alignments.

The rotor's axle is extended on one side out of the motor enclosure and an application device may be attached to it. There is a wealth of rotary machinery applications in our kitchen, driven by electrical motors. It is noteworthy that their mechanical activity also relies on rotation. All blenders, grinders and mixers do blend, grind and mix their respective subject of operation with rotating motion.

The same principle applies to the hand tools in our life. The widely used hand drills (including the rather dreaded tool of dentists) and the circular saw both attest to the unique applicability of the rotary motion to remove material from a block. Even for tools where a linear motion is needed, like in the case of a surface sanding tool, the power is provided by an electric motor and transfer of the rotating motion to linear is done via a gear.

Industrial machine shops bring another class of rotary equipment. The lathe and the milling machines are nowadays elaborate, computer controlled complexes of multiple, adjustable axes of rotation. The result is automated machining of very delicate shapes with high precision and very brief times. Rotary machinery dominate our manufacturing world.

Another consequence of the rotation, centrifugal force is also utilized in various rotary machines. The simple household washing machine's speedy cycle employs the centrifugal force to push the water out of the clothes being washed. In more advanced centrifuges scientists are separating components of mixed materials of different densities, hence inertia. Then there are centrifuges used for nuclear material purification that may again become a main staple of energy supply in the future of humankind.

A reverse manifestation of the electromagnetic phenomenon, the electric generator, turns mechanical energy provided by some external source into electrical current. The external source could be from any natural resource, such as the wind or moving water.

Wind mills, known for centuries, use a set of blades attached to a shaft that is freely rotating under the pressure of the wind.

Its childhood toy version, the paper stripes on a small stick we all ran around with, is the proof of the concept working well. In its use in the middle ages, preceding the concept of electricity, the rotational energy obtained on the shaft from the wind was translated into other mechanical uses, such as pulverizing grains or lifting water with a water wheel.

In modern wind mills the rotating axle of the wheel is extended into a generator enclosure with wiring on the stationary part. The end of the rotor axle is now the place for the magnets that induce the electrical current in the wiring. There are wind farms in many areas of the world, ranging from the California desert regions to the Scandinavian sea shores.

The concept of scooping up water with a water wheel brings us to the other natural energy source: moving water. In a reverse application of the ancient tool, the water wheel may also be simply subjected to the force of the water allowing the wheel to freely rotate along with the axle it is mounted on. The other end of the axle was driving grain grinding stones already in the middle ages, a device transcending all the early human societies with agricultural culture.

In the modern arrangements, water is gathered up beyond a dam and the water wheels' successors, the water turbines are built into tunnels. The force of the water in the tunnels rotate the turbine blades mounted on rotating axles. On the generator side of the turbine the electricity is produced by a reverse use of the electric motor phenomenon komplained above.

Both of these devices, wind mills and water turbines, use abundantly available natural mechanical energy to generate expensive electrical energy. Humankind has an ever increasing energy requirement for the world and needs clean ways of generating it. Hence the rotational energy, obtained by harnessing the power in the motion of the water and air, is still the most promising energy source for the future.

Most rotating objects are circular or spherical, but that is far from being a necessity. Propellers, the most important tools of mobility of mankind after the invention of the wheel, are not circular in themselves in stationary position. They have specially shaped blades, and several of them are attached to a cylindrical hub. They are arranged at even angular distances.

The profile of the blade is somewhat different whether it is aimed for moving (in) air or moving (in) water, but we'll ignore that in our focus on the rotating phenomenon. In most propellers, be it on a ship, an airplane or a windmill, the number of blades is two to four.

The rotating motion of the propeller blade through some medium, like water or air, will produce a thrust on the hub. The thrust will move whatever is attached to the hub (a boat or an airplane) forward in the particular medium. It is Archimedes' genial screw concept at work here again, as the propeller will also have a spiral movement in the medium when moving ahead.

19

GALACTIC ROTATION

It is rather peculiar that the planets of our solar system rotate very close to a common plane. Why is that? The question confounded Newton when he realized the fact that gravity keeps the planets moving around the Sun. His explanation, based on the prevailing creationist wisdom, was that God must have placed the planets into a plane.

It took the genial effort of Laplace again to provide a scientific explanation in a monumental five volume work titled Celestial Mechanics. In the book published in 1799, literally in the last days of the most tumultuous century of France's history, Laplace was finally able to explain why all celestial objects revolve in the same direction and largely in the Sun's equatorial plane. He described a whirling solar atmosphere as the origin of the arrangement.

Laplace hypothesized that the effect of rotation to a cloud of gas with solid particles resulted in the observed system. If the rotational speed and the density are high enough, the particles of the cloud will gravitationally attract each other and coalesce into bigger chunks. As this process continues, the cloud will rotate faster and faster, somewhat similarly to the ice skater's spinning being accelerated by the skater closing the arms around the body.

The originally spherical cloud will become more and more flattened and the largest chunks of the cloud (the future planets) will orbit in the disk. Since the originally irregularly shaped parts will have inertia forces acting on them, they will start to rotate themselves. That rotation will further shape those chunks into spheres and planets will be formed, as shown on the figure.

The birth of planets

This hypothesis is now commonly accepted. When astronomers looked at young stars, they found that many of them have a surrounding being disk of gas and particles. Obviously, we cannot really monitor one of them becoming a system of planets as the celestial time scales are measured in millions of years, but it appears that the process is valid.

The 19th century brought Napoleon back into power, albeit for a short time culminating in his final defeat at Waterloo. During this time Laplace dedicated a volume of his work on celestial mechanics to Napoleon, his former student. As the story goes, after briefly leafing through the book Napoleon complained about seeing no mention of God in it, perhaps the first such comment of the historical discourse surrounding the topic. In his later years Laplace became a strong supporter of the Bourbon royal family and Charles X granted him nobility and the Marquis de Laplace title in 1827.

Since the process proposed by Laplace appears to be operational in millions of places in the universe, it is likely that many other solar systems of our kind may have been formed during the billenea of our past. The inevitable conclusion is that extraterrestrial intelligence may also have developed. Some of those may even have visited us before, but this topic is far beyond our rotational history focus.

We have just established that rotation played an instrumental role in the forming of our solar system. Let us now move on to the host of our solar system, the Milky Way galaxy and see the instrumental role of rotation on that scale. Milky Way is also rather flat, disk-like and has an assortment of interesting spiral arms as visible on the cover image. This shape implies that there is some rotation involved there as well.

The Milky Way galaxy is home to another 100 billions suns and solar systems, like ours. The two closest solar systems in our galactic neighborhood are Sirius and Alpha-Centauri. Our galaxy is roughly 140,000 light years in diameter and its thickest part in the center, the bulge, is about 20,000 light years in diameter. The galaxy also contains billions of meteoroids and asteroids, along with hundreds of star clusters.

In the beginning of the last century modern methods of radio wave and X-ray observations arrived in astronomy, certainly more powerful than Galileo's hand-made telescope some 300 years earlier. Jan Oort was born in the Friesland area of the Netherlands and studied at the famous Groningen University. His main area of interest was astronomy, especially distant stars and he pioneered the detection of radio waves arriving from them.

Observing the radio waves from many distant Milky Way objects, in 1925 Oort realized that some stars are lagging behind the Sun in their motion while others are bypassing it. This, he recognized, was similar to the motion of the inner planets vs. the outer planets in our solar system. He considered this to be the direct proof of the rotation of the Milky Way, but it was not accepted immediately.

There is now irrefutable evidence that the Milky Way is rotating. There is an eerie resemblance of the Milky Way picture on the cover to that of the hurricanes on Earth. It is quite likely that our solar system may also be controlled by the forces of rotation we seem to have understood by now, the centrifugal and Coriolis forces.

Further observations of Milky Way also established the presence of three major spiral arms, called the Sagittarius arm, the Orion arm and the Perseus arm, in order of their distance from the center of our galaxy, the last being the farthest out. There are several minor arms and some that seem detached from their original major arm.

So where are we in this rotating galactic system? Oort calculated that the galactic center would be about 30,000 light years away from us. We are on a rotational path around the center of the Milky Way on one of those spiral arms, specifically the partial arm of the Orion arm, called the Orion spur. We are also about 15 light years off of the main plane of the disk.

Consequently we rotate around the center of the galaxy at an approximate speed of 234 kilometers per second. At this speed it takes us about 250 million years to complete a rotation, mere minutes compared to the age of the universe, presently assumed to be about 15 billion years. Our solar system is estimated to be about 4.6 billion years old, hence our Sun may have already done about 18 full circles.

It is hypothesized by some astronomers that our solar system was actually born about ten percent farther out, some 33,000 light years from the center and 200 light years off the main plane of the Milky Way. It appears that we are moving toward the central gravity providing object of the Milky Way (the black hole we conjecture it to be now) and on a spiral rotation pattern like some galactic hurricane under the Coriolis force. The apparent clockwise rotation indicated by the spiral arms may even mean that we are on the "southern hemisphere" of the galaxy, using the notation rather figuratively.

There are of course other galaxies, besides ours. The universe is estimated to contain about 100 billion galaxies. Our Milky

Way is actually a member of a group of about thirty galaxies, called the Local Group. Our closest neighbors are the Magellan cloud and the Andromeda cloud, the latter is a subject of some intrigue in the next chapter. It is safe to assume that the rotational phenomenon played an instrumental role in their formation as well. This begs the question of the applicability of the rotational principle to the intergalactic space and eventually the universe.

After all, if the Milky Way's shape resembles a hurricane, that is the movement of air in the atmosphere of a rotating Earth, then that shape may also imply that it is moving in a rotating universe.

20

ROTATING UNIVERSE?

From the days of Ptolemy and Copernicus, the bigger picture of the universe was represented by the so-called fixed stars in the sky. The ancient astronomers attempted to establish our position in the universe in reference to those known constellations. This became the Zodiac, the basis for a medieval branch of astronomy called astrology. It is nowadays considered a pseudo-science at best, but just like the flat Earth belief, it still has followers.

Astrology was based on affixing certain characteristics to the perceived appearance of twelve major galactic constellations, Sagittarius, Scorpio, Libra, Virgo, Leo, Cancer, Gemini, Taurus, Aries, Pisces, Aquarius and Capricorn. Their alignment was associated with certain periods of the year and people born in those periods supposedly inherited the generic personality characteristics of their sign. The horoscope generation was a custom made application of these astrological principles to a particular person's time of birth.

It was also used to interpret the cycles of our life, from birth through adulthood and to death, and in turn giving birth to other lives with their own cycles. There is even an out-shoot of this, called synchronicity, that associates certain celestial events with actual Earthly events. This leads into the realm of the highly speculative, so we will end this train of thought.

Let us return to the astronomical side and reconsider the precession cycle of Earth's axis introduced in an earlier chapter. The 25,760 year precession represents a very long cycle in the sky, shown on the figure below, where the number 2000 represents the calendar year 2000, our present celestial north in the sky. The total cycle is shown by going back 10,000 years and going forward about 16,000 years, their sum totaling the precession cycle time.

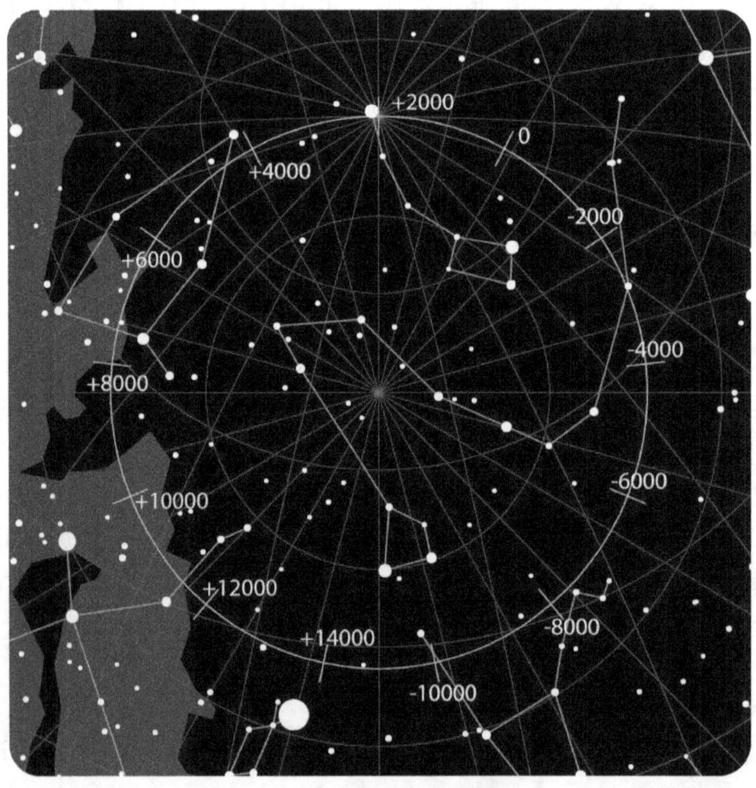

The precession cycle

The size of the spots in the precession cycle picture indicate the relative luminosity of the stars. The rather bright star to the left of the 2000 year mark is Polaris that is our present North Star. The same concept is applicable to the southern hemisphere. The polar direction of the southern hemisphere, or the South Star is in the Sigma constellation.

The brightest star on the bottom of the picture is Vega in the constellation named Lyra. Vega will become our North Star in about 14,000 years. Polaris will become our North Star again in about 25,760 years or in the year 27,800 AD, assuming that there will not be any change in the current calendar system. We can't be sure whether there will be humans to observe that time period in the life of our planet, but the planet by all celestial indications will still be alighted by the Sun.

Since this rotation in the sky is too slow for a human life time observation, even for human cultures, we do not truly comprehend its effect on our life. The turns of this rotational cycle bring apparent ice ages about ten thousand years apart and some natural warming periods between them, subject of extreme cultural and political upheaval in our society today. We may, however, only have a childish tempest in the universal teapot in that regard.

A giant step in humankind's quest of understanding the universe was taken by Edwin Hubble in 1924. Hubble was born in 1884 on a Missouri farm and as his famous predecessor, Galilei, he also first studied the topic of his father's choice. In Hubble's case this was law, but his private interest was in astronomy. He became one of the first recipients of a Rhodes scholarship and went to Oxford in 1910, where his astronomy interest deepened.

His scientific breakthrough came at the Mount Wilson observatory in Pasadena. After diligent observations for long nights on the cool mountain top, he recognized the fact that the Andromeda Cloud was just too far away to be a part of the Milky Way and it is a galaxy on its own right. This finding destroyed the prevailing single galaxy view of the time. He showed that the universe was full of galaxies and he also conjectured that they are apparently speeding away from each other. The universe is expanding!

This expanding universe does not preclude the possibility of the galaxies rotating, a fact we found very plausible in connection with the Milky Way. Then the physics interpretation in the rotating frame of reference of the universe may also include a certain manifestation of the Earthly phenomena of the centrifugal, Coriolis and gravitational forces on a truly universal scale.

But the rotating galaxies seem to have an apparent discrepancy between the masses of galaxies conjectured based on their luminosity and the masses required to account for their motion. An explanation, presently accepted by scientists in the know, states that the discrepancy is due to material that has no luminosity hence invisible, the so-called dark matter. As many radically new ideas in astronomy, it is only grudgingly accepted. While it approximately explains the newest observations of certain stars moving within other galaxies, we have no concrete proof for its existence.

The orderly rotating vision in the sky is sometimes disturbed. Intriguingly the Andromeda galaxy, Hubble's subject of observation appears to be on a collision course with the Milky Way. That is of course in the realm of possibilities, after all there are loose planetary objects in our own solar system as well. Since those defy the general rotational scheme of things, there could be rogue galaxies in the inter-galactic space too.

As with Earth's rotation around Sun in the solar system, we need a reference system with respect to which the rotating phenomenon is measured. However, if our universe is infinite as put forth by many scientists, then there should not be anything outside of it. Then there would be no external reference system, hence the universe could not be rotating.

This scenario, of the universe being all encompassing, immobile and infinite, requires an explanation of the origin of the material in this unique universe, the creation of the gas-dust clouds in various places of the universe, and the central masses controlling the rotation of the galaxies. It is presently explained, although not universally accepted, that the origin is an infinitely small, infinitely dense primordial material cloud preceding an explosion called the big bang. This hypothesis appears to be supported by Hubble's expanding universe observation, but it still requires a big leap of faith to believe that, since we do not know physical laws that would govern the phenomenon.

This issue of the creation of the original material is not to be confused with the formation of the stars and planets. Many of those phenomena are now indisputable and many still are hypothetical. The issue is much deeper and gets as philosophical as physical. After all, if the universe was finite and there was a hyper-universe around it, our universe could be rotating with respect to an axis specified in that hyper-universe.

There are ongoing attempts trying to quantify the rotation of the universe. It is presently conjectured to be an extremely slow rotation completing a turn in about 60,000 billion years. This so-called "universal vorticity" translates into about 10^{-12} degrees per year angular velocity. The question is, with respect to what? This is a question with profound implications.

We seem to have a very good understanding of the workings of our solar system that may be carried to our galaxy and to the known universe. But when reaching beyond, we require an inquisitive mind not satisfied with the standard explanation of the big bang. The glaring singularity in the beginning of that hypothesis leaves room for other ideas and a potential proof of the rotation of the universe may reveal some hidden facts regarding its creation.

If the axis of rotation of our universe is located in a hyper-universe, then the phenomenal effects of rotation could work on an even higher scale. Then a new chapter of the history of rotation may be written.

EPILOGUE

We arrived at the end of our historical journey. We saw that the gradual understanding of the rotational phenomenon as it applies to our solar system is one of the greatest epics of human history. It is filled with cultural division and religious opposition, all to be overcome by physical evidence and the strong personae of the scientists involved.

The scale of the rotational phenomenon is also very important, especially when the rotating motion of extraterrestrial objects of our galaxy is observed. The size of a human being is on the order of a meter, most people are in the range of 1.5-2.0 meters (between five to seven feet) in height. Our cosmic rotational horizon, the distance we can see at the moment (not visually of course), is about 10^{26} meters, something that we cannot put into everyday words in our units. One can say that humans expanded their horizon from the size of their being with about 26 orders of magnitude. Astonishing, almost incomprehensible, but it is true.

What about narrowing the horizon? Looking inside, into our bodies, at the materials we and the world are comprised of, is another interesting journey one can take. That road also goes through various rotational phenomena, starting from the first detailed, so-called planetary model of the atoms. That model posited a nucleus of the atom surrounded by electrons on circular patterns in an eerie resemblance to our solar system.

This was later improved by recognizing that the electrons' orbits are matching certain energy states and when they change orbits, some new particles are ejected. The concept of various rays such as alpha and gamma emerged. Even more interesting is the fact that it was also found that the orbiting electrons, besides the angular momentum of their orbit around the nucleus (like Earth's angular momentum in its orbit around the Sun), also have a spin (again like Earth does around its own axis of rotation). The similarities are even more striking if we consider that this phenomenon is now played out in the scale of about 10^{-16} meters.

Very likely it does not end here, as the continually discovered sub-particles prove. Since the traversing of the 26 orders of magnitude outward took millennia, but the traversing the 16 orders inward happened mainly in the last one hundred years or so, there is obviously much more to learn in that direction. We only have rudimentary understanding of the physics of the subatomic scale. Perhaps in another hundred years we'll even out our external-internal playing field in God's playground.